PRACTICAL MINERALOGY, ASSAYING AND MINING;

WITH

A DESCRIPTION OF THE USEFUL MINERALS,

AND

INSTRUCTIONS FOR ASSAYING AND MINING

ACCORDING TO

THE SIMPLEST METHODS.

BY
FREDERICK OVERMAN,
MINING ENGINEER.
AUTHOR OF "MANUFACTURE OF IRON," AND OTHER WORKS OF APPLIED SCIENCES.

Sixth Edition.

PHILADELPHIA:
LINDSAY & BLAKISTON.
1863.

STEREOTYPED BY J. FAGAN.

INTRODUCTION.

The United States abound in valuable minerals, which are strewed all over the surface of the country, and imbedded in its soil. These minerals, particularly those which are most useful in the arts, are not so generally known as they should be; and those who usually become possessed of them, seem to be the most ignorant of their worth and practical application.

It has always been the desire of the author, to place before the public the characteristics and uses of minerals, in a popular style, and clothed with a popular language, so that all who can read may have an opportunity of fully understanding this interesting subject

For this reason, he has endeavoured to avoid, as far as possible, the use of any scientific and technical terms, as having a tendency to embarrass, rather than to enlighten the reader.

The subject has been divided into three parts: MINERALOGY, or, a descritpion of the appearance of minerals, and of the localities in which they have been, or may be found; ASSAYING, or an investigation of the value of minerals, by means which are within the reach of every one; and an essay on PRACTICAL MINING, in its most simple forms.

THE AUTHOR.

PHILADA. Mar. 1, 1851.

TABLE OF CONTENTS.

PART I.

Distribution of Minerals .... Page 13
Granite 14
Metamorphic Rock 16
Stratified Rock 18
Tertiary Formation 19
Volcanic Rock 20
Alluvium 21
General Remarks 22
Geographical Distribution 24
Origin of Minerals 25
Position of Minerals 25
Depth of Veins 26
Faults 27

Description of Minerals 28
Iron and Iron Ore 28
Native Iron 28
Brown Hematite 28
Red Iron Ore 29
Magnetic Iron Ore 30
Carbonate of Iron 32
Sparry Ore 34
Iron Pyrites 35

Copper 38
Native Copper 38
Sulphuret of Copper 38
Copper Pyrites 39
Red Oxide of Copper 40
Black Oxide of Copper 41
Hydrosilicate of Copper 41
Carbonate of Copper 41
Phosphate and Chloride of Copper 41
Variety of Copper Ore 41
Lead 42
Galena 42
Carbonate of Lead 44
Phosphate of Lead 44
Gold 46
Native Gold 46
Silver 56
Native Silver 56
Sulphuret of Silver 57
Chloride of Silver 59
Antimonial Silver 59
Antimonial Sulphuret of Silver 60
Platinum, Iridium, &c. 61
Mercury 62
Native Mercury 62
Sulphuret of Mercury 62
Bituminous Sulphuret of Mercury 62
Chromium 64
Chromic Iron 64
Manganese 64
Black Manganese 65
Zinc 66
Native Zinc 66
Zinc-blende 66
Red Zinc Ore 67
Calamine 67

Tin 68
- Tin Stone 68
- Tin Pyrites 69

Alum-Slate 70
Alum-Stone 70
Amber 71
Asphaltum 71
Asbestos 72
Chalk 72
- White Chalk 72
- Black Chalk 73
- Red Chalk 73
- French Chalk 73

Clay 73
- Loam 74
- Potters' Clay 75
- Kaolin 76
- Slate Clay 77

Coal 79
- Mineral Charcoal 79
- Anthracite 79
- Bituminous Coal 80
- Brown Coal 83

Diamonds 85
Fullers' Earth 86
Garnets 87
Graphite 87
Grindstones 88
- Whet-Slate 88
- Emery 89
- Flint 89

Heavy Spar 90
Limestones 90
- Compact Limestone 91
- Magnesian Limestone 91
- Lithographic Stone 92

Shelly Limestone 92
Gypsum 93
Marble 94
Marl 95
Ochre 96
Salt 96
Sand 97
Sandstone 98
Slate 98
Roofing Slate 98
Soapstone 99
Sulphur 99
Tripoli 100
Turf 100

PART II.

Assaying of Minerals 101
Examination of Water 102
Rain Water 103
Spring Water 103
Saline Water 105
Turbid Water 105
Solution of Salt 106
Examination of Solid Matter 107
Minerals imbedded in Sand 107
Minerals from Veins 110
White Mineral 111
Yellow " 113
Red " 113
Brown " 114
Black " 115
The Lens 117
Examination of Gold Ores 118
Roasting the Mineral 119
Pounding the Mineral 120
Distinctions between Roasted Minerals 126

Roasted Iron 126
Copper 126
Zinc 127
Lead 127
Antimony 127
Tin 127
Quicksilver 128
Gold 129
Assay of Ore by Smelting 129
Crucibles 130
Cupels 132
Smelting 133
Smelting Furnaces 134
Assay of Iron Ore 135
Assay of Copper Ores 140
Preparation of Black Flux 141
Assay of Copper Ore at Swansea, in Wales 142
Refining the Assay 148
Assay of Lead Ore 150
Assay of Gold Ore 155
Assay of Silver Ore 158
Refining Lead to obtain the Gold and Silver 161
Formation of a large Cupel 161
Formation of a Muffle 163
Process of Cupelling 163
To separate Gold from Silver 165
Assay of the Ore of Mercury 167
Assay of Tin Ore 169
Assay of Zinc Ore 171
Assay of Chrome Ore 172

PART III.

MINES 174
Hunting for Minerals 176
Searching by Trench 178

Searching by Shaft 179
Searching by Drift 179
Searching by Boring 179
Miners' Tools 181
Blasting Tools 182
Blasting 185
Mining 190
Dead Work 190
Excavation of a Gallery or Drift 192
Timbering a Drift 193
Size and Kinds of Timber 195
Drainage 196
Shafts 199
Timbering of a Shaft 200
Extraction of Minerals 203
Working in the Open Air 203
Digging of Clay 206
" of Valuable Sand 207
" of Marl 207
" of Iron Ore 208
" of Limestone 208
" of Gold 209
" of Anthracite Coal 210
Subterranean Workings 210
Horizontal Veins 210
Working of horizontal Veins, less than six feet thick . . 216
" of horizontal Veins, more than six feet thick . 214
" of horizontal Veins, beneath a Plain 214
" of Mineral Deposits of great thickness 215
" of Inclined Veins 215
" of heavy masses of Minerals 219
Reflections 221
Ventilation of Mines 222
PRACTICAL REMARKS 226

PART I.

MINERALOGY.

DISTRIBUTION OF MINERALS.

THE useful minerals are not distributed over the earth irregularly; there is a system in these distributions, the knowledge of which furnishes us with a rule by which we decide whether a particular kind of mineral may be expected to be found in a certain spot. The signs by which we judge whether a particular mineral is present at the place in question, are the general characteristics of the rock in that locality. The knowledge of the relation of particular minerals to the general character of rocks constitutes the science of Geology; and the knowledge of the character of minerals, that of Mineralogy.

It is not our object to penetrate into the science of Geology; but it will be useful to give a general idea of the positions in which minerals are found, and of their relations to each other.

GRANITE.

The history of the formation of rocks has been divided into certain periods; and it is generally agreed that *granite* is the oldest of the rocks. For this reason it is called primitive rock. Granite is a close, compact rock, composed of fragments of other rock or stony matter. These are so firmly cemented together, that the whole forms but one solid mass, without the slightest indication of pores or fissures. The matter of which granite is composed is often found to be in the form of small crystals, seldom or never assuming the shape of round grains. It is found of all shades and colours, from a bright white to deep black, often in the same block. The crystals are, in many instances, not more than one-twelfth of an inch in diameter; but they are also found of the size of one inch, and even larger. If a certain colour of the composing crystals predominates, the granitic rock appears to be of that colour, either grey, reddish, greenish, or bluish; shades of yellow and crimson are also perceptible. Granite rock is particularly characterized by the absence of all stratification, or any indication of parallel joints; the rock is uniformly compact in all directions. It is of great hardness and strength, and of everlasting durability. It takes a fine polish; but, on account of its component matter scaling off in leaves, like mica,—which latter is sometimes found to be one of its elements,—it does not take or retain a solid surface.

Granite is not very extensively found in the United States. It occurs chiefly in the States of Maine, Massachusetts, New Hampshire, Connecticut, New York, along the Lakes, and sometimes in the Mississippi Valley.

Granite is an expensive material, but it is the most durable that can be employed for building purposes. Extensive use is made of it in the cities of Boston, New York, and Philadelphia, as a building-stone, among which the Quincy granite, from Massachusetts, may be considered a fine quality. The ancient Egyptians made very extensive use of granite for monuments, which, after the lapse of thousands of years, are found to be as fresh in appearance as if they had just come from the chisel of the artist. In modern times, the sculptors of Europe also employ granite for monuments. This rock forms a first-rate material for paving and macadamizing roads. It is unsurpassed for paving-stones, of which some of the granite-paved streets of the Atlantic cities are evidences. Granite may be used as hearth-stones in blast-furnaces for smelting iron and other metals. It is particularly qualified for making moulds in which copper or brass plates are cast, which are afterwards to be rolled.

Granitic rock is frequently interspersed with more or less vertical crevices or veins, which are filled with matter foreign to the rock itself, and form lodes and veins of ores or other minerals. We may expect to find in these veins, ores of tin, iron, copper, lead, cobalt, silver, a few other metallic ores, and anthracite coal. We find also, in such veins, feldspar, kaolin, quartz, in beautiful crystals, plumbago, or black lead, garnets, heavy spar, calcareous spar, fluor-spar, and fragments of rock of various kinds. We cannot expect to find bituminous coal in granite; nor anthracite coal, except as a curiosity. We do not find gold, platinum, iridium, rhodium, and similar metals, nor sulphurous ores of tin, nickel, cobalt, or mercury; nor sulphur and sulphurets in such quantities as to justify their extraction.

METAMORPHIC ROCK.

The rock of this formation, also called transition rock, is the second in age. To this class belong a great variety of minerals; as gneiss, mica-slate, clay-slate, limestone, and other minerals, in rocks covering tracks of great extent and at a great depth. The rock of this class is characterized by a partial and sometimes by a decided stratification. It does not exactly belong either to the compact or the stratified variety.

It is beyond our province and object to give a particular description of the various classes of rock belonging to this series, as the mere enumeration of the names would occupy several pages of this book. Metamorphic rock, which often assumes the appearance of granite, pudding-stone, or stratified rock, is particularly distinguished by its close grain and strength from the rocks of the secondary formation, and by its stratification from granite and volcanic rocks.

Transition rock covers the greater part of the New England States, New York, western New Jersey, eastern Pennsylvania, Maryland to the Alleghanies, middle Virginia, parts of North Carolina, South Carolina, Georgia, Alabama, Missouri, Arkansas, and all the States west of the Mississippi. In this rock, which is the most extensive in the United States, we find, and may expect to find, gold, in Virginia, North and South Carolina, Georgia, Alabama, New Mexico, California, Utah, and Oregon. We also find silver in this rock in Vermont, Virginia, the Carolinas, Georgia, Arkansas, New Mexico, and California. Platinum and the platinum metals are also found along with the gold. Lead is found in this rock in almost every State, particularly in Missouri and Arkansas. Iron also is found everywhere in this formation, and generally the best quality of iron ores. Besides these, there are found ores of zinc, antimony, arsenic, nickel, cobalt, tin, manganese, and in fact almost every kind of ore. In Pennsylvania, Virginia, and North Carolina, heavy deposits of plumbago are met with; and anthracite coal is almost exclusively found in this formation. It is the home of metallic sulphurets. It generally forms good building-material, if not too hard to be worked. The sandstones of this period are exceedingly well qualified for building purposes, of which an example may be seen in the brown Connecticut sandstone, so extensively used for ornamental architecture. We find roofing-slate, and extensive tracts of limestone, in the rock of this period, in New York, New Jersey, Virginia, Ohio, Missouri, Illinois, and other States. The limestone is generally magnesian; that is, it contains magnesia as well as lime. The mineral veins of this period run either parallel with the stratification, as the gold veins of the southern States do; or traverse the strata in more or less inclined angles; or form isolated elliptical masses or lodes, as the iron, zinc, and lead ores generally do. This rock formation is the most productive in useful minerals; and where a faint indication of something valuable is discovered, it is generally worth the trouble to follow and dig after it.

STRATIFIED ROCK.

The rock of this period, also called secondary formation, or coal formation, is decidedly stratified. We easily follow the various layers of minerals, which are distinctly parallel to each other. The layers or seams are sometimes almost horizontal, as is the case in the western coal basin. The inclination of the strata of the Pittsburgh coal veins is almost imperceptible. We can follow the same stratum for hundreds of miles, without its disappearing below ground, or running over the tops of the hills. In other places the strata are more or less inclined; the coal strata of Alleghany county, Maryland, form a canoe, sunk between two high ridges of mountains. In Illinois and Missouri we find the strata undulating, gently rising and falling.

The rock of this period is the *coal-bearing* rock par excellence. We find here the richest and most extensive layers of mineral coal, all of the bituminous kind, or soft coal. The great western coal field covers western Pennsylvania, parts of Ohio, Illinois, western Virginia, Kentucky, Tennessee, Alabama, Indiana, Missouri, and other States. The Maryland coal formation is about forty miles long, and eight miles wide; the Pittsburgh field is over five hundred miles long, and from two hundred to three hundred miles wide. There are coal deposits of this kind in the transition rocks of eastern Virginia, North Carolina, California, Oregon, Nova Scotia, and Canada. In this formation we are to look chiefly for soft mineral coal, iron ore, limestone, and salt. It contains none of the precious metals, no lead, no copper, nor any metal except iron and manganese, the latter not available. There is no plumbago or anthracite, no heavy spar or fluor-spar. We can neither find nor expect to find any thing useful but sandstone, (which, however, is an inferior building-stone,) fire-clay, limestone, iron ore, and salt in the form of brine, which is found from one hundred to one thousand feet below the surface. Coal is almost everywhere found where this kind of rock appears.

TERTIARY FORMATION.

Stratified rock of a later period is frequently called tertiary formation. In this rock we find plenty of shells, and fossil remains of animals and plants. This rock is not rich in minerals. It often contains a species of mineral coal not much valued, and sprinklings of iron ore of no consequence. The only mineral of importance in this series is green sand or marl, which is extensively deposited in heavy beds all along the Atlantic coast, from New York to Florida. This marl, containing the elements for stimulating the growth of plants, is extensively used in New Jersey and Maryland, and to some extent in Virginia, for improving the land. This formation, extensive as it is, offers but little inducement to search for minerals; and those which are found are generally of an inferior quality. It affords no building material of any consequence, and the small beds of shelly limestone contained in it are not of much importance. In many instances, a fine quality of potter's clay is found in it; but this is generally of such a composition as to be too fusible for strong and fine ware. It frequently affords fine sand for ordinary glass.

VOLCANIC ROCK.

The rocks belonging to this class are often found to be perfectly vitrified, and of a glassy appearance, as basalt and some kinds of lava. The first is found in columns, grouped together in isolated mountains, or imbedded in other volcanic rocks. The latter appears in compact masses, and is often very porous. To this class belongs a remarkable rock, called trap-rock. This is generally very hard, tough, and crystalline in its fracture. We find it also soft and brittle. This rock is found all along the Atlantic coast, intruding from below in the form of both small and large dikes, imbedded in the transition rock. It is very extensively found around the Lakes and in the Rocky Mountains. This rock is characterized by the presence of copper and heavy layers of iron ore. The native copper and silver of Lake Superior is imbedded in it; and in New Jersey, Pennsylvania, and the southern States, we also find copper ores in or near the trap-rock, Iron ore is found imbedded in this rock in almost all the States of the Union. Of the volcanic rocks, this is the most valuable as a matrix of useful minerals. The other rocks of this class are of less importance.

All the rocks of this formation are hard, and, when broken into small pieces, if not partially decomposed, emit a sound like crockery, when struck with a piece of wood, or against one another. This rock furnishes the very best material for macadamizing roads, and some of it forms the most durable building-stone; but it is very difficult to work.

Of useful minerals, we find in these rocks copper, gold, silver, iron, sulphur, alum-slate, arsenic, lead, pumice-stone, and other substances.

ALLUVIUM.

Alluvium is a term used to designate the most recent deposits of matter. It comprises deposits of gravel, sand, loam, and clay; and frequently contains, or is entirely composed of, animal remains. It is found along the sea-coast, the lakes, rivers, and rivulets, and forms their banks and bottoms. Alluvium is generally loose ground or fragments of rock. Most of the valleys, swamps, and prairies are alluvial matter.

We find the following useful minerals in these deposits of earthy matter: iron ore, (bog ore only,) potters'-clay, fire-clay, gravel, brown coal, peat or turf, and some other minerals which will be mentioned hereafter.

GENERAL REMARKS.

Minerals, particularly the metallic ores, are chiefly distributed over and below the surface of the earth in layers or masses; in lodes, or large veins, running parallel with or traversing the general course of the stratification of the rock; in nests or pockets; in nodules, which are concretions or accumulations of minerals of small extent; and in small veins, which are either branches of heavy veins, or traversing larger veins, ramifying the rock in all directions.

The transition rock is the domain of metalliferous deposits. Here we find more variety and abundance than in any other geological formation. In this rock the best qualities of ores are also found. The secondary rocks, or those which contain bituminous coal, do not furnish so great a variety of minerals as the transition rocks, and many kinds are not found at all in them. In the latter there is an absence of gold, silver, copper, antimony, and other substances. It furnishes chiefly stone-coal, iron, limestone, and often a little lead or zinc, but these latter only in a very limited quantity.

Mineral substances, particularly those which are used for the production of metals, — the workable or useful ores, — are but few in number. They are generally oxides, that is, combinations of metal with oxygen; sulphurets, or metals combined with sulphur; and carbonates, or metallic oxides combined with carbonic acid. Combinations of metals with other substances than these, are rare.

Minerals of the above description frequently form large bodies of pure ore by themselves, buried beneath the surface in the rock. In most cases, however, these ores are blended with foreign matter, as lumps or grains of quartz, lime, and other substances, which are mechanically mixed or in chemical combination with it. In some instances, the main body of the layer or vein is rocky matter, and the ore imbedded in that substance. The mineral deposits are often found in such heavy masses as to admit of their extraction without any admixture of rock; in other cases, rock and mineral must be raised together, as they cannot be separated. In some instances, the veins are too small for the entrance of the miner, and are consequently of no practical value.

GEOGRAPHICAL DISTRIBUTION.

The geographical distribution of minerals is very distinctly marked in the United States. The ridge of the Alleghany Mountains divides the surface of the country into transition rocks on the east side and secondary rocks on the west. In the first, or between the Atlantic and the Alleghanies, we find chiefly the minerals and rocks belonging to the transition series; and in the second, or between the Alleghanies and the Mississippi river, we find chiefly the minerals belonging to the secondary rock. Both slopes of the Rocky Mountains are of a uniform character, and the minerals on each side are those belonging to the east slope of the Alleghanies. Precious metals abound, to all appearance, more on the western than on the eastern side of that mountain chain.

The extent of mineral veins is as different as the character of the minerals themselves. In the western coal fields, particularly in that of Pittsburgh, we may trace the same vein of coal for hundreds of miles, as well as veins of iron ore and limestone. There are also small veins of coal and ores, of but a few miles in extent, in the same basin. The veins of gold ore in Virginia and North Carolina can be traced in a continuous belt for more than five hundred miles, running parallel with the Alleghany ridge. Immense beds of magnetic iron ore are found in the State of New York; also veins of peroxide of iron of uncommon magnitude in the Mississippi Valley. Layers of useful minerals are found in almost every State of the Union, in such abundance and extent as to render the raising of them most easy.

ORIGIN OF MINERALS.

The origin of the minerals, and their form as veins or layers, may be considered as the result of infiltration from the surface, to which class many of the iron and copper ores belong; or the deposits have been formed in the bottom of a sea, as those of the coal measures; or the minerals are injected from below, raised by the power of internal heat, to which class the gold and silver ores of Virginia and North Carolina, and the native copper of Lake Superior, belong. The first class generally consists of wedges decreasing with the depth; the second of spheroidal masses, and the third of wedges increasing with the depth. The first class of veins is the most deceptive, and cannot be depended upon; the second may be measured by its appearance on the surface, or by sinking shafts into it; the third class may be depended upon as improving with the depth.

POSITION OF MINERALS.

Masses and veins of minerals are not always found in a horizontal position. The stratification of the great western coal field is almost horizontal; and as the mineral veins of this region are parallel to the stratification, these also are of course nearly horizontal. Except in the bituminous coal region, we find but few horizontal veins of minerals. The veins of Virginia and North Carolina run parallel with the stratification of the rock, and also with the layers of it; but the layers are in many cases almost vertical, and generally inclined not less than 60°. The inclination or dip of a vein is measured from the horizontal plane; if, therefore, a vein dips but 10°, it is nearly horizontal; and if it dips 80°, it is almost vertical. The dip of a vein is not always the same in its various parts, particularly if the plane of the vein is not parallel to the plane of the stratification.

DEPTH OF VEINS.

It is almost impossible to determine the depth of those veins which are not horizontal, or nearly so, so long as their utmost depth has not been reached by actual working. The mining operations of the United States are of so recent origin, and there is so great an abundance of minerals on the surface, that there has been hitherto neither time nor necessity for deep mining. There are but very few mines more than three hundred feet in depth; we cannot therefore say how deep the mineral veins descend in the United States. There are mines in Germany of the depth of twenty-six hundred feet; in Mexico, sixteen hundred and fifty feet. The tin and copper mines of Cornwall, England, are eighteen hundred feet deep; and the silver mines of Norway, Saxony, and Hungary are of equal depth.

FAULTS.

Veins of minerals are frequently found to be disturbed in their regular course, either by other mineral veins or by dead rock. Such disturbances, called *faults*, slips, or slides, appear in every kind of vein; they are caused by matter which has penetrated the crevices of the rock after the main vein was formed. The mass of a vein is often found to divide itself into various small veins, which, at certain distances or at greater depths, reunite. Such faults, whether consisting of mineral or dead veins, are often perplexing to the practical workman; but the scientific miner is never at a loss in such cases.

DESCRIPTION OF MINERALS.

In the description of minerals we shall follow their relative value, as it is represented in the quantity deposited in the United States. We shall commence with the metallic ores, and follow with the other minerals in their order.

IRON AND IRON ORES.

Native iron is a mere curiosity, of no practical value. It is found in Connecticut.

Brown hematite—brown oxide of iron; brown iron-stone; pipe ore; bog ore.—This is found of almost all shades of colour, and under the most varying forms. It is characterized by its powder when rubbed, or its streak, which are always yellow. We find this ore of all shades of yellow, brown, and black. Its lustre varies from the dulness of loam to the resinous brilliancy of pitch. Its compact or solid varieties are generally granulated, but are frequently found of fibrous texture and of a silky lustre, the fibres being from a lively brown to jet-black. It is mostly opake, but often transmits light through thin scales and at its corners, in which it appears to be blood-red. It sometimes appears in the form of hollow nodules, which are often of a black velvety appearance on the inner surface. This ore is so extensively distributed, and appears under so many different forms, that a description of it is very difficult. The most certain mode by which it may be distinguished is, to reduce it to a powder; if this is yellow, the ore belongs to the variety under consideration.

The scientific term for this ore is hydrated oxide of iron. Its chemical composition, if pure, is peroxide of iron, with from 13 to 18 per cent. of water. In its purest condition it never contains more than 60 per cent. of iron.

This kind of ore is the most profusely distributed over the United Strtes. It is found in heavy beds near and in the anthracite region of Pennsylvania, and in the valleys of the western coal formation, where it forms the outcrops of the veins of argillaceous carbonates of iron. It occurs in Massachusetts, Connecticut, Tennessee, Kentucky, and Alabama, sometimes in veins thirty feet thick. The iron ores found in the bogs of New York, Michigan, Ohio, and Illinois, belong to this variety. It is found in almost every State of the Union, and is the most generally distributed of the iron ores. It is best qualified for pig-metal; the impure varieties not being well adapted to the manufacture of bar-iron.

Red iron ore—red hematite; iron-glance; specular iron ore.—Until within the last few years, there were some doubts as to the quantity of this ore being greater than that of magnetic ore in the United States; but we now assert that it is found more frequently and in larger masses than the latter. The appearance of this ore varies from a dull brownish red, like reddle, to the lustre and colour of polished steel or plumbago. Its powder often feels greasy, like plumbago, but it is always red when rubbed upon white paper or on a white porcelain plate. Some kinds of this ore, like that of the Missouri iron mountain, are compact, and possess the colour and lustre of steel; other ores of the same kind are found in crystals, in the form of fine leaves or cubes, and of the colour and lustre of black lead, as is the case in New York, New Jersey, Pennsylvania, Arkansas, and other States. A heavy body of this ore, of the red variety, has been found in Wisconsin and Michigan. Smaller veins are found in most of the States of the Union.

The chemical composition of this ore is iron and oxygen; and, if pure, it may contain about 70 per cent. of metallic iron. It is frequently adulterated with clay or siliceous matter, and is often found to contain but from 10 to 20 per cent. of iron. Some kinds of red clay ore, though of an intensely red colour, contain but 5 per cent. of metal. This ore is not attracted by the magnet. If not too largely mixed with foreign matter, it forms one of the best and cheapest iron ores for the smelter. The quality of iron made of it is always found to be soft and strong. It is particularly qualified for the production of heavy wrought iron.

Magnetic iron ore—loadstone; black oxide of iron.—Large beds and veins of this ore are found in the United States, particularly on the west side of Lake Champlain, in Essex county, New York, and in the

States of New Jersey, Pennsylvania, and Ohio. It is also found in the New England States. This ore is generally bluish-black, and sometimes pitch-black. It is of a metallic lustre, and exceedingly hard. It is found in compact, solid masses, and also in crystalline grains, from a large size down to the form of fine black sand. The compact and the crystalline varieties are frequently found in the same vein. This ore is characterized by always forming a black powder when rubbed upon a white body. If it contains impurities, its powder is more or less grey. It is exceedingly sensitive to the magnet, and is attracted by it. When it occurs in large pieces, it attracts iron, and imparts magnetism to it when rubbed over it. This ore is said to belong to the primitive rock formations; but in the United States we find it chiefly in the metamorphic rocks of Pennsylvania, New Jersey, and New York. It forms the main body of the iron ore in Sweden.

If pure, this ore is better qualified for making strong iron than any other, provided it be not spoiled in the smelting operation. In Jersey City, an excellent cast-steel is manufactured of the iron derived from this kind of ore, found in the western part of the State of New York. Spring-steel and file-steel are also extensively made of it in New Jersey.

This is the richest of the iron ores. The compact ore of Lake Champlain, which is nearly pure, contains 70 per cent. of metallic iron in 100 parts of ore. There are varieties which contain but from 20 to 25 parts of metal; these are conglomerates, in which the crystals of ore are imbedded in a cement of clay, silex, and often lime. Magnetic ore is frequently adulterated with foreign matter injurious to iron, as silex, copper, arsenic, titanium, and particularly sulphur; the latter often in large visible quantities, in the form of crystals of yellow pyrites. This is the case with most of the richest veins of magnetic ore in New Jersey and Pennsylvania.

Carbonate of iron.—This species comprises two varieties. The first of these, termed the compact, or argillaceous ore, is chiefly found on the western side of the Alleghanies. The other variety is the sparry ore, found on the eastern slope of that mountain chain. The first is the most extensively distributed in this country, and for that reason we shall speak of it first.

The compact carbonate of iron—clay iron-stone; argillaceous ore—is chiefly found in the western bituminous coal formation. It is there deposited in veins of more or less extent and thickness. Some of these veins are more than fifty miles in length and eight feet in thickness; others are so small as not to be workable. The general form in which this ore is found is that of a flattened spheroidal body, from the size of a pea to a mass of sufficient bulk to weigh half a ton. These balls form either a continuous vein, in which one is laid beside and above the other, and the spaces between them are filled with clay, or the balls are separated, sometimes many feet or yards, and imbedded in slate. We find this ore also in continuous veins, in a compact form, resembling limestone. All this kind of ore, when discovered near the surface, at the outcrop of the vein, is found to be decomposed, has lost its carbonic acid, and is converted into brown or yellow hematite (hydrated oxide). By following these veins or outcrops of veins, we always find the ore in the interior of the rock to be the compact carbonate. The finest qualities of this ore are found near Baltimore, which is not in the coal region. We also find it in the Frostburg coal region, in Maryland, and in almost all the western coal deposits. The colour of this ore is sometimes white, but generally of a dirty grey, a yellowish-brown, or of a faint brick-red appearance. There are oxidized veins of this ore in the western coal-fields along the Alleghany and Ohio rivers, of fine quality, making a superior iron. Most of it adheres to the tongue like clay, and emits an odour like that of clay when breathed upon. All our bituminous coal formations contain this ore, which is not so generally the case in Europe.

This carbonate of iron generally contains from 20 to 33 per cent. of metallic iron; seldom more than 36 per cent. Its composition is protoxide of iron, carbonic acid, clay, silex, lime, and often magnesia; and in most cases it contains manganese, which is often found in the centre of a decomposed ball, forming a black lump. The balls of this ore, when in the progress of decomposition, form shells of hydrated oxide of iron, which are distinguished by different colour, and may be separated in the same manner as the different coats of an onion. In the centre of such a ball, we often meet with an undecomposed core, where the ore is in its original state, presenting the appearance of limestone.

Carbonate of iron certainly constitutes the largest quantity of iron ore in the United States; but, on account of its being of a difficult and tedious treatment in preparing it for smelting, it is very little used. The Baltimore variety furnishes a superior metal, very much esteemed in the Atlantic markets for making wrought iron. Some of the best iron in the Western States is manufactured of this kind of ore, decomposed; as that of Hanging Rock, Ohio.

Sparry ore—sparry carbonate; also called spathic ore.—This is the second variety of the native carbonates of iron. In Europe, it forms the first quality and quantity of the two; but we cannot say that it does so in this country. There it is also called steel ore; but that term does not apply here. The largest quantities of this ore are found in Vermont, Connecticut, and New York. It is also found in smaller veins in all the New England States, in New Jersey, Pennsylvania, Virginia, North Carolina, and the States around the Lakes.

In almost all instances where this ore occurs, it is adulterated with sulphur, and in some cases with copper, which detracts seriously from its practical value. It is not much used. The colour of this ore is in most cases white, varying to yellowish-brown and dark-brown. Its texture in the fresh fracture is always decidedly crystalline, and of a silky lustre. It is not attracted by the magnet,—which is also the case with the compact variety before described;—but if either kind be slightly heated, it is attracted by the magnetic steel.

This ore is frequently found to form the main mass of a vein in which other valuable ores are present; and in this respect it is a guide to detect ores which would not otherwise be found. In North Carolina, it forms the bulk of a vein of gold ore, where it is accompanied by quartz, iron and copper pyrites, and a large quantity of gold. It associates with all kinds of metallic ores, changing the character of a vein from one kind of ore to another.

The foregoing enumeration constitutes the bulk of useful iron ores. There are some few ferruginous minerals, which are used in the manufacture of iron; but they do not constitute iron ores proper, and may be considered as fluxes. Among these substances are ferruginous slate, shale, and clay-slate, which contains iron, red marl, and green marl. These minerals contain but from 5 to 10 per cent. of iron. Any mineral which does not contain at least 20 per cent. of iron, is not considered an iron ore.

Iron pyrites—sulphuret of iron; in some places simply called sulphur.—This we do not consider as iron ore; but it is a species of mineral of great value in some parts of the country. There are two different kinds of iron pyrites: the one is yellow, of a brass or gold colour; the other is white, of a silvery lustre. The chemical composition of both is nearly alike. Each of them contains more than half its weight of sulphur; the other part is metallic iron. This mineral is frequently confounded with more valuable substances, by those who are not expert metallurgists, on account of its great lustre, bright colour, and hardness. It is easily distinguished from any other mineral; for the slightest heat drives off sulphur, the suffocating smell of which at once proves its character. This mineral is exceedingly hard: it strikes fire with steel.

Sulphuret of iron is very extensively distributed all over the United States, and accompanies almost every description of mineral. It is found in all geological formations, in primitive rock as well as in alluvial gravel. In the coal regions it is distributed in small veins, leaves, or crystals, incorporated with the coal, and depreciating its value. Where the coal is slaty, and contains much sulphuret, — or, as it is commonly called, sulphur, — it is used for making copperas, or alum, provided the slate contains but little coal. Along the Ohio river, extensive use is made of the sulphurets in this way. In other parts of the United States, not much attention is paid to these minerals.

Iron pyrites are of little value in themselves; but, as a matrix of other metals, namely, as the bearers of gold and silver, they deserve more attention than they have hitherto received. All iron pyrites contain gold, and often silver, from which rule only those of the coal formation are excepted. The extensive gold deposits of the Southern States constitute virtually a belt, or accumulation of veins, of iron pyrites. The gold had its seat originally in the pyrites, which, when decomposed, liberate the gold, and it appears in a metallic state. The pyrites are the matrix of the gold. The veins of gold ore in those regions are, and have been, veins of pyrites, decomposed at the surface to a certain depth; below that decomposition, the veins are essentially formed of pyrites, and at a greater depth, entirely so. The pyriteous slate of these regions contains gold in most cases, if the pyrites are perfectly decomposed. It does not follow from these statements that all the sulphurets of this kind contain so much precious metal as to make them worth working. It is not impossible that the heavy pyriteous veins of Pennsylvania and New Jersey contain gold.

Iron pyrites and copper pyrites are not easily distinguished from each other. The first, however, is of a decidedly crystalline form, the latter not so; the first is very hard, the latter does not strike fire with steel. The colour of the iron pyrites varies from a pure silvery white or golden yellow to red; the copper pyrites are of all the colours of the rainbow.

COPPER.

Native Copper, and copper ores are, after iron, the most important of the minerals. Native copper is found in large quantities, in regular veins, in the State of Wisconsin, near Lake Superior. The heavy masses of copper in these places are imbedded in volcanic rock, and small veins ramify it in all directions. It occurs in bodies of almost every size, from small grains to masses weighing ten tons and upwards. This native copper is frequently found to be mixed with silver in distinct fibres, the latter not being alloyed with the copper. Native copper is found in almost every vein of copper ore; it has been found in those of New Jersey, and also in those of Pennsylvania.

Sulphuret of copper.—The ores from which the most copper is smelted are the sulphurets of copper. There are two kinds of ore of this variety; the one is called grey sulphuret of copper, and other copper pyrites; the latter generally contains iron in admixture. The grey sulphuret is a compact ore; its surface is dull, and it is of the colour of lead, or an iron-grey; it also occurs of a faint red colour, if taken near the surface of the ground. It melts easily, if a small splinter of it be held in the flame of a candle. It may be cut with a hard and sharp knife. If the ore is pure, it contains 78 parts of copper, and 18 parts of sulphur. It is found in New Jersey; and a heavy vein, of fine quality, occurs in the copper region of Lake Superior. It is always found in veins of copper ore, forming a part of their mineral contents.

Copper pyrites resemble iron pyrites very much, but may be easily distinguished from the latter by their iridescence, or lively rainbow colours. This ore is always accompanied by iron pyrites, the latter often decidedly predominating. It contains, if pure, from 70 to 80 per cent. of copper; but it is not found in its pure state. The best specimens contain from 12 to 20 per cent. of iron. These ores are found in all the transition and metamorphic rocks of the United States, particularly in the New England States, all along the Atlantic coast, around the Lakes, and near the Rocky Mountains. It is the most common copper ore, varying in yield from 2 and 3 per cent. of metal to 40 per cent. Ores yielding a smaller proportion still are found, but they are valueless. Copper pyrites are very abundant, but are frequently so much mixed with other matter as to make mining unprofitable. Iron is the metal most commonly associated with them. They are mixed with magnetic iron in eastern Pennsylvania, New Jersey, and New York; with sparry carbonate of iron in Vermont and North Carolina; and with iron pyrites everywhere. In Virginia, North Carolina, and the gold region, we find this ore in all the metallic deposits. It accompanies the gold-bearing iron pyrites in all instances, and is considered a good indication of richness. It is also found along with sparry iron, galena, sulphuret of zinc, and, in fact, all the gold and silver ores of that district. There are very few instances in which the ore in these veins will pay for the labour and expense of extraction and transportation; the amount of metal rarely exceeding 4 or 5 per cent. It is said that large quantities of copper pyrites are found in Tennessee.

Considering the extensive distribution of this ore over the United States, its great utility to the smelt-works, and its general demand, it is to be regretted that the mining of it is not carried on more extensively. The smelt-works of the Atlantic cities pay $2.50 for each per cent. of copper in the ore. As this brings a ton of 10 per cent. ore to $25, and as ore containing this proportion of metal is abundant, and railroads and canals abound almost everywhere, it certainly ought to be a good business to dig copper ores. If we consider that in Germany copper ores which, in most instances, contain but 1 per cent. of copper, and a little silver, are extracted and smelted to advantage, it cannot be considered difficult here, where the veins are frequently found to be heavy, to work mines which furnish ores containing 5 or 6 per cent. of copper.

The foregoing are the most generally distributed copper ores. We find others, but their quantity is comparatively small. The presence of the following forms of this ore gives us an assurance of success in deep mining:—

Red oxide of copper, distinguished from oxide of

iron by a lively red colour, more brilliant than the latter.

Black oxide of copper, which is velvet-black, often inclined to blue or brown.

Hydrosilicate of copper, or *siliceous oxide of copper*.—This is generally the green ore, found in most cases at the outcrop of veins of copper ore. It has a bright green colour and resinous lustre; and when freshly broken, its fracture is like that of glass.

Carbonate of copper, or *malachite*.—This is generally of a blue colour; but it is also found of all shades, from dark blue to light green. Specimens of this variety are found in every copper vein, particularly at the outcrop; but they are better fitted to occupy a place in a cabinet of curiosities than to serve any useful purpose.

Phosphates and *chlorides of copper* are also found. They are both green, and form no regular copper ore.

There is a great variety of copper ores. They are of various shades of colour, from a brilliant red to velvet-black; of a beautiful green, and sky-blue. They are all distinguished from other ores by their bright colour, and their power of imparting that colour to other substances. There is scarcely any variety of copper ore which does not betray the presence of that metal by a green film, particularly if it has been exposed to the influence of the atmosphere. Copper pyrites, if exposed to the air for a short time, exhibit a film of blue crystals of sulphate of copper.

LEAD.

This is the next metal that demands our consideration.

Native lead is a mineral curiosity, of no practical value whatever. It is said to have been found in its metallic state; but that is of little consequence, because it can have been discovered only in small quantities, and because it costs but little to smelt it from its ores. The most important ore of this class is

Galena, or sulphuret of lead. This ore has the lustre and colour of polished metallic lead. It is always grey, without a shade of any other colour; but its powder, when finely rubbed, is black. It is always found in a crystalline form, the crystals being cubes, often composed of square plates, and frequently so small as only to be detected by the aid of a lens. In other instances, the cubes, or the plates which form the cubes, are more than one inch square. Galena is composed of metallic lead and sulphur; it contains, if pure, 86 per cent. of lead and 13 per cent. of sulphur. It is very heavy, and equal to metallic iron in specific gravity. It is, indeed, the heaviest of all metallic ores.

Galena is very extensively distributed over the United States. It is found almost everywhere, except in the bituminous coal region. The most extensive deposits are in the States of Missouri, Illinois, and Arkansas; between the Blue Ridge and the Alleghanies, in Virginia; and in smaller quantities in all the States of the Union. It is extensively worked on the upper Mississippi.

Some varieties of galena contain silver to a large amount; but in that case the ore is generally mixed with other minerals. The Missouri galena contains but little silver; not enough to pay for separating it from the lead; but the Arkansas galena — which, unhappily, is sent to England to be smelted — is said to be very rich in silver. The latter appears to be a very pure and beautifully crystallized mineral, consisting almost entirely of lead. It seems to form an exception to the rule, that pure lead ores are poor in silver. Galena containing silver is very extensively distributed over Virginia and North Carolina, and, in fact, over the whole gold region; but it is not so generally in use as it deserves to be. The difficulty of smelting this ore profitably appears to be in the way of its more general application. A large smelting establishment has been erected upon a vein of this ore in North Carolina, and appears to be in a prosperous condition. The ore of this description in the gold region is of a bluish tinge, often inclining to black, and contains an accumulation of small crystals, which may be distinctly recognised in it. It is a compound of the sulphurets of zinc, lead, iron, copper, tin, silver, and gold. The galena in this ore amounts to from 5 to 10 per cent. of the bulk; and each ton of lead smelted from it contains from 180 to 200 ounces of silver and gold, and frequently more than that. This may be considered a rich silver ore, and will pay a handsome profit to a properly managed smelting concern. Any galena which yields 50 ounces of silver to a ton of ore, is considered a silver ore. This amount of silver alone will pay for refining the lead.

Carbonate of lead.—This is a lead ore of frequent occurrence; but it rarely forms a vein of itself. It accompanies other lead ores, in the form of soft white concretions, as a powder, or in crystals. The crystals of carbonate of lead are generally flat, and transparent like glass. The most beautiful specimens of this kind of ore have been found at the Washington mine, North Carolina. When the crystals, or the earthy kinds of this ore, are kept in a room where stone-coal or coal-gas is burned, or where men or animals breathe, it gradually turns grey, and at last black; forming black sulphuret of lead, by the absorption of sulphuretted hydrogen. The clear crystals of this ore possess the double refracting power; that is, they show two images of any object viewed through them. They are very soft, and easily scratched or broken.

Phosphate of lead.—This variety is not so common as the foregoing, but occurs quite frequently in all lead districts. It is an indication of the presence of lead, for it is chiefly found at the outcrop of a vein. It is, however, too rarely found to be of much practical value. It generally occurs in and near the gold regions of the Southern States. Its colour is greenish; and if inclined to yellow, it contains arsenic and phosphorus. We find it, like the carbonate, or native white lead, in the form of a fine powder, in concretions, and in crystals.

There are, besides the above, quite a variety of minerals containing lead, but they are of little interest as ores. There are oxides, sulphates, chlorides, arseniates, &c., which are of little interest to the practical man. The lead ores appear of almost every colour, from the brightest white in the sulphates, to the deepest black in the sulphurets. All these various compositions of lead, if exposed to the air in a room, turn black, except the sulphate, which remains white. The colour of these lead ores is generally not distinct, but is of a dirty, earthy character, or becomes so in the atmosphere. They are all, however, characterized by forming a fine powder when rubbed, which possesses the adhesive property in an eminent degree, and are in this respect superior to any other mineral. Finely powdered lead ore, particularly galena, is used for glazing earthenware.

The galena found in limestone formations, or accompanied by lime, is generally poor in silver. The richest and most numerous beds of ore are found in and near limestone rocks. The lead ores of siliceous formations, particularly those found in slates, are generally rich in precious metal; and it may be said that the lead ores of the oldest rocks, as, for example, those from granite, are the richest.

There is no doubt that there is more lead beneath the surface of the earth in the United States, than we are aware of; however, the price of lead is so low in the principal markets, that the raising of lead ores cannot be considered a profitable business, unless the body of ore is very large and can be raised cheaply. Still, lead ore may be very useful, if the amount of silver in it is sufficient to pay for extraction, smelting, and refining. Such lead ores are found, and may still be found, if searched for, in the metamorphic rocks along the Atlantic coast, near the Lakes, and on both slopes of the Rocky Mountains. Lead ores in the limestone region will not pay to be worked in that way; but those from the slaty rocks will pay.

GOLD.

It may be wondered why I did not put this article at the head of my list, because it is certainly, at the present time, the most valuable mineral in the United States, representing a greater value than all others combined. As an excuse, I must confess that I have no confidence either in the profitableness of the mines in California, or in their permanency. If California slacks a little in its productions, or the iron market improves a little, the balance will turn in favour of iron, as to the value of the yearly production. Minerals which contain native gold, or the regions where gold is washed from the alluvial gravel, are very plentiful; and at the present time more productive in this country than in any other part of the world.

Native gold. — California stands pre-eminent in

the production of this precious metal. The gold in these regions is found in its native state, in small grains, in spangles, in crystals so small as to be almost invisible to the naked eye, and also in lumps of ten and twenty pounds weight. These grains of gold are often found to be imbedded in masses of quartz; at other times, they are mechanically enclosed by quartz; but in most cases the grains are pure gold, alloyed with a little silver. The latter admixture diminishes the value of California gold about 15 or 20 per cent. The gold of California is almost exclusively found in the alluvial sands and gravel on the banks of the rivers, and in the valleys traversed by mountain streams. The implement used for washing out the gold is most simple, consisting of an iron or tin pan. When this vessel is filled with sand, immersed in water, and shaken, the gold sinks to the bottom, and the sand, clay, and gravel flow off with the water, or are taken out by hand. It cannot be doubted that a large quantity of gold is lost in this way; but it appears that more perfect machinery is not applicable, because the deposits of gold are spread over a large surface of country, and transportation of machinery is expensive.

There are gold-bearing localities in Virginia and North Carolina, which, if not equal to those of California at present, will be of greater importance in future, and, I predict, more sure and lasting. There is very little gold extracted from alluvial sands and gravel in North Carolina at present. There were formerly alluvial gold diggings in the States mentioned, which yielded as well as the best California placers. Such diggings furnished, some years ago, large quantities of gold. The abundant yield, however, did not last long; the rich deposits were soon exhausted, and the poorer localities did not pay. Most of the gold in Virginia and North Carolina is at present derived from veins by mining. The gold-bearing rock is chiefly a talcose slate; that is, a slate resembling soapstone, but which does not feel so greasy. This slate is red, and ferruginous at the surface. At a greater depth, it is filled with small crystals of iron pyrites, which are decomposed near the surface, and appear as peroxide of iron, which colours the slate brown, and, in a few instances, yellow. This slate is of various grades of hardness. It is generally softer in Virginia than in North Carolina. In the latter State it is found to be very hard in those places where the ground or rock has been under the influence of more heat than at other places. This slate is a metamorphic rock, and runs in a regular belt, parallel with the Alleghany mountain chain. Its length is at least five hundred miles, extending from Maryland to the south-western extremity of North Carolina. The width of this gold-bearing belt is, in its broadest part, from twenty to twenty-five miles, which is often contracted to two or three miles. Within this belt, the various veins of gold-bearing slate are distributed. Those parts of the vein which are richest in gold, are characterized by small veins of quartz, running parallel with the slate. Where this quartz is wanting, not much gold is to be expected. The talcose gold-bearing slate of Gold Hill, North Carolina, is particularly distinguished in this respect; and it may be said that the ores from that mine are the richest of the whole gold region. The direction of the veins is parallel to the general course of the rocky strata or formation; that is, from north-east to south-west; and their inclination, which is also parallel to that of the strata, is from 45° to 90°. This belt of talcose slate extends farther north-east, through Maryland, Pennsylvania, and New Jersey; but it is changed in its composition. It appears in this extension as mica-slate, and ceases to contain gold. Farther south-west than North Carolina, it changes into feldspar and its relative rocks. The gold in this south-western district is found imbedded in heavy veins of quartz, which appear frequently in this rock, being parallel with its stratification. These veins of quartz are often twenty feet thick and upwards; they are pyriteous, and contain iron, copper, and sulphurets of lead, which are found to be rich in precious metal. The gold-bearing belt, which can be traced in a southerly direction through South Carolina, Georgia, and Alabama, sinks beneath the Mississippi river, and probably rises again to the surface near the Rocky Mountains.

The gold found in the slate of Virginia and North Carolina occurs in exceedingly small grains, often so fine as to be not only invisible to the naked eye, but undiscernible even by the assistance of a strong lens. This is the case even when the ores are worth three or four dollars per bushel. Some veins of the slate region contain coarse gold, in grains as large as the head of a pin, and even larger. These are generally found in veins of quartz, in which the pyrites are concentrated into larger masses. Where the pyrites are disseminated in fine crystals through the mass of the rock, the gold is found to be very fine. In the fresh pyrites the gold is invisible, even if, after separation, it appears to be coarse. By natural or artificial decomposition, the gold becomes visible; the pyrites are converted into oxide of iron, and, by the aid of a lens, the gold may be detected, imbedded in the oxide of iron.

Another form in which native gold is found, is in quartz, in which it is imbedded. Solid white quartz, both in veins and in crystals, is found, in which the gold occurs in spangles, plates, grains, and also in perfectly developed crystals. Quartz of this description is met with in Virginia; more perfect specimens occur in North Carolina, and still better in Georgia; but the best quality is found in California. The gold in this form is rather a curiosity than a valuable mineral, for no regular veins of such gold-bearing quartz have yet been found, which promised good results.

Gold never appears in solid veins; it is always disseminated through the mass of the rock, in some places more dense than in others. There are localities in the gold region of the Southern States, where every piece of rock, every handful of soil, contains more or less of the precious metal. Gold is never found in secondary strata or the coal regions. We may look in vain for it on the western slope of the Alleghanies; it cannot be there. Its origin appears to be in primitive rock, or granite; but it is most abundantly found in the trap-rocks, or those of igneous origin. The geological formation of California is of this character; but that of the gold region of the Southern States is not. In the gold-bearing strata of these States we find trap-rock frequently intruding, but it is not the matrix of the gold. Greenstone-porphyry, syenite, and gneiss, appear to be the primary sources of gold. These are also found in dikes and veins in the gold regions. The immediate matrix of gold in these regions is evidently the pyrites, which, however, may be a secondary enclosure. This opinion is supported by the fact that the richest gold ores are found near a vein of trap-rock or other igneous strata.

It is the general impression, when gold is found in the bottom of a stream or near its banks, or in alluvial soil, that a vein of gold ore must exist somewhere above that place, where the gold is found. This impression, however plausible, is fallacious. We find gold in grains disseminated through granite, and also meet with gold washings in the alluvial deposits of granitic mountains; still, there are no gold-bearing veins found in this rock. Transition rocks contain also spangles and grains of gold, but more commonly veins, in which the gold is associated with quartz or calcspar, but most frequently with pyrites. The primary source of gold is evidently in granite or its associate rocks; and the presence of coarse gold in the igneous or volcanic rocks may be attributed to a coagulation of the small particles in its primary sources. These particles belong to the mass of the rock, and not to veins located in it. The veins found in these rocks must necessarily be of an origin secondary to the rock itself; and, as such, have been either infiltrated from above or injected from below. The infiltrated veins can only consist of quartz, feldspar, calcspar, and similar infiltrations, which may include gold accidentally washed into the fissures of the rock. Pyrites and all other sulphurets of metals are injections from below: these cannot crystallize from a watery solution. These sulphurets have been driven into the crevices of the rock, either in the form of vapours, — which is most probable, — or have been injected in masses by pressure from below. It appears, then, from the foregoing remarks, that we are not justified in supposing veins of gold to lie in the neighbourhood of the alluvial deposits where that metal is found. On the contrary, such a supposition appears quite improbable; for the gold from pyriteous veins—the true source of that metal—is generally very fine, and is most likely to be carried off by the waters. It is an observed fact, that where the gold in the alluvial washings is coarse, the chances are, that there is no regular vein of gold ore near. Coarse gold is mostly found to have been distributed through the mass of the rock. Where the streams contain more gold after heavy rains and freshets, it is an indication of there being no veins whence the gold is derived: the abrasion of the rock furnishes the metal. Where the gold in an alluvial deposit is found in a stratum, it is an indication of there being no vein. A severe winter or a heavy freshet is the cause of the formation of this stratum. A vein would furnish a regular supply, not form a stratum. Gold is never carried far from its original resting-place; therefore a vein cannot be found at any considerable distance from the alluvial washings. There is no prospect of finding it where the river ceases to carry it, either above or below.

The quantity of gold derived from this source may be very promising at first, but its future continuance is quite doubtful. There are circumstances which indicate the origin of the California gold to be in the masses of the rock; and it may be well for our Californian friends to pay some attention to these facts.

The next deposit and source of gold may be found in the infiltrated veins. Gold enclosed in crystallized quartz is evidently derived from alluvial soil, which has been washed into the crevices of the rock, and afterwards covered with quartz in solution; and to this result the heat of a volcanic region has no doubt greatly contributed. Silex is easily soluble in pure hot water, but is precipitated from it as soon as it comes in contact with any other matter, or when cooled. The crevices of the feldspathic rock of North Carolina are chiefly filled with crystalline quartz, which in many instances contains gold. This quartz is evidently the result of infiltration; and all the veins of this kind must be uncertain in their duration and extent.

The veins injected from below are a third source of gold. To these belong the pyriteous veins, and, as far as their decomposition is concerned, the ferruginous veins. Whether the gold in these veins is in a metallic form, and has been evaporated in that state; or whether the gold was raised and condensed along with other metals and sulphurets, is a question of no importance. It may be asserted as a fact, that all native sulphurets, particularly all the sulphurets of iron, contain gold. It does not follow from this that all pyrites contain sufficient gold to pay for its extraction. As sulphurets cannot possibly penetrate any rock but from below, we may naturally conclude that the heaviest body of such kind of ore must necessarily lie deep in the earth. This conclusion is supported and confirmed by practice; for all pyriteous veins are invariably found to improve in quality and quantity with the depth. This circumstance speaks very favourably for the gold formation of the Southern States. We have here a belt of gold ores of unparalleled extent, immense width, and undoubtedly reaching to the primitive rock, which, on an average, cannot be less than two thousand feet deep. Here is a mass of precious metal, enclosed in the rock, which cannot be exhausted for ages; and in this respect the region in question is the most important of all the known gold deposits, California not excepted.

From the foregoing considerations we may conclude that the sources of gold decide the value of a mining district. The gold derived from the abrasion of rocks, where the metal is promiscuously disseminated, is the cheapest, if in sufficient quantity, and coarse, as is the case in California. The first miners who arrive at a virgin deposit may make a fortune in a short time; but, when the gold at that spot is exhausted, there is no vein to fall back upon. It requires a series of years, perhaps ages, to accumulate another heavy deposit of metal. If the washings in such districts are not carried on too extensively, a regular yearly supply may be depended upon; the crop being in proportion to the quantity of rain which has fallen during that period.

The yield of gold from its ores is very variable. In alluvial deposits, a fortune may sometimes be seen in a bucketful of sand. California and North Carolina have each furnished lumps of native gold worth upwards of six thousand dollars. The gold ores of the Southern States — that is, the ferruginous slate, either oxidized or pyriteous — yield from ten cents to one hundred dollars per bushel or one hundred pounds of ore. Oxidized ore which yields fifteen cents' worth of gold per bushel, pays very well for extracting. A profitable business is done with such ore in North Carolina. Ores which yield but ten cents' worth to the bushel are worked to advantage. Most of the mines, however, yield from twenty-five to thirty cents' worth of gold per hundred pounds or bushel of ore. At Gold Hill, North Carolina, the ore yields, on an average, one dollar per bushel.

If the inexhaustible quantity of the gold ores in the southern states is considered, and their capability of paying at least the expenses of extraction and profits on investments, it may be said that the poor ores of that region are ultimately the most profitable to the miner. In crushing, washing, and amalgamating these ores, a large portion of gold is wasted, which in poor ores amounts to 50 per cent. Ores which yield twenty cents' worth of gold by amalgamation, yield forty cents by smelting them. The undecomposed ores are troublesome to work. The way adapted to obtain all or most of the gold is, to amalgamate the ore at different times, with intervals for decomposition by exposure to the air.

SILVER.

Native silver occurs in various forms; and it is often difficult to decide by sight whether a mineral is pure, or contains silver in admixture. It is found in all mines where silver ores occur, in the regular form of crystals, but chiefly in irregular grains and formless aggregations. It appears in the native copper of Lake Superior, ramifying the copper in all directions, in the form of fine threads of pure silver. Native silver has been found in the Washington mine, North Carolina, in large masses, in irregular concretions, in crystals, and in bundles of fine fibres, like hairs, growing from a common centre. Metallic silver is also found in the New England States in small quantities. Silver has a great affinity for sulphur, which soon blackens its bright surface. For this reason we find most of the native silver in black masses, imbedded in the silver ores, filling fissures in a vein, or appearing as a black vegetation in cavities or on the surface of a vein. Metallic silver is found in all mines where silver ores are found, or where the ores of another metal are impregnated with silver. Most of the silver in the United States is derived from gold. All the gold brought to the mint from the mines, contains some silver, varying in amount from 1 to 15 per cent., and upwards. All native and manufactured silver contains gold, copper, iron, or arsenic. These metals have a great affinity for silver, and cannot be entirely separated from it in the smelting operation. In the United States, but little silver is smelted from the ores. The only establishment of any extent is the one located at the Washington mine, above mentioned. Silver ores are found in abundance in this country, but little use is made of them.

Sulphuret of silver — silver-glance. — This is the

most common of the silver ores. We find it in the form of crystals, hairs, and needles, or like wire twisted into nets; in plates, and in shapeless masses. This ore is opake, or of a dark-grey colour. It is malleable, and easily cut with a knife, like lead. It is not elastic, like metallic silver. The clean cut looks like metallic lead, but is soon covered with a film of various colours. This ore, in its pure form, contains 87 per cent. of silver, and 13 per cent. of sulphur. It is easily smelted, and yields metallic silver with but little trouble. Sulphuret of silver, and all the silver ores, are found in rocks of all ages, except in the coal region; and always accompany the ores of copper, lead, antimony, gold, arsenic, and others, along with quartz, calcspar, heavy spar, manganese, pyrites, and other minerals. It is a remarkable fact, that silver occurs more abundantly where mineral veins cross or meet each other, than in other places, or in the finer ramifications of a vein.

Sulphuret of silver is the chief source from which silver is smelted in this country. It is abundant in the gold region of the Southern States, where it appears in heavy veins, associated with other metallic ores. Of all the deposits of this kind, the ores of the often-mentioned Washington mine are the only ones which are smelted to any considerable extent. This is so much the more remarkable, as there is a large body of such ore in Virginia, as well as in North Carolina; and it is said that a similar ore appears in Tennessee. The silver ores of the gold region are imbedded in a grey, blue, or brownish-black mineral, which is composed of from 30 to 50 per cent. of sulphuret of zinc, from 5 to 10 per cent. of galena, in small crystals, some iron pyrites, copper pyrites, sulphuret of tin, and, in some cases, a little arsenic. The amount of silver in these ores is variable; it amounts to from twelve to sixty ounces in each ton of crude ore, of which the ounce is worth nearly two dollars. The high price paid for silver from these regions, is on account of the large amount of gold with which it is alloyed. These ores are exceedingly rich in gold. Some of them yield it by simply being pounded and washed.

Horn-silver—chloride of silver.—This ore is not so generally found as the above sulphuret, but it appears in almost every place where silver is found. It occurs chiefly at the outcrop of veins, along with native silver or sulphuret of silver. Chloride of silver is a horny substance, so soft as to be cut by the finger-nail. Its colour is often grey, and it sometimes shows all the colours of mother-of-pearl. These colours darken, if the ore is exposed to light for some time. This ore is also found of a uniform green colour.

One variety of chloride of silver is called buttermilk ore. In this case it is mixed with foreign minerals, and with gold and copper. The greater portion of it, however, is clay.

Antimonial silver.—This is the richest of the silver ores, but not so frequently found as others. It has not been observed in the old States of the Union, It is found in Mexico and Central America, and may exist in the new acquisitions of the Union, New Mexico, California, Utah, or Oregon. Antimonial silver is a crystallized ore, of a white or yellowish-blue colour, hard, and very brittle. It resembles arsenical iron very much, but is easily distinguished from that ore by its crystals being longer, and not quite so hard. The ore, when fresh from the mine, is white; but it is soon tarnished by a yellow film, and gradually becomes grey, blue, and at last dark-grey or black. This ore, when pure, contains 80 per cent. of silver and 20 per cent. of antimony.

Antimonial sulphuret of silver—red silver; ruby-blende.—This is a valuable silver ore, not yet discovered in North America. There is, however, little doubt of its being found. Its colour embraces all the shades of red, and is sometimes of an iron-grey. It is rarely found of any other colour. The lustre of this ore is remarkable, being metallic, and in many instances as brilliant as that of a diamond. It is found wherever other silver ores are found, and may be expected in the mineral veins of primitive, transition, and metamorphic rocks. It is accompanied by or associated with antimonial and arsenical ores, lead, cobalt, nickel, copper and iron pyrites, along with quartz, lime, heavy spar, fluorspar, and other minerals.

There are other varieties of silver ore, but their appearance is very rare.

An important source of silver ore appears to be in California. We have expressed our doubts as to the permanency of the lucrativeness of the gold diggings in those regions; but, according to all accounts, there appears to be a regular and well-developed system of silver ore mines, which will be of great service after the gold is so far exhausted as to afford but usual wages to the miner. These veins of silver ore appear to be favourably located, fuel for smelting it being convenient to the mines.

PLATINUM; IRIDIUM; OSMIUM; RHODIUM; PALLADIUM.

These are called the platinum metals, because they always appear together or alloyed. Platinum has been found in the gold diggings of Virginia, North Carolina, Georgia, and California; but not in such quantities as to be of any importance. These metals are as valuable as gold; and some of them are sold at even higher prices than that metal. They are chiefly found wherever gold occurs, and mostly or exclusively in alluvial gravel and sand. Platinum appears in flattened grains, of a greyish or lead colour, resembling tarnished steel. In its ordinary state, it is as heavy as gold; and, if pure, even heavier.

MERCURY—QUICKSILVER.

Native Mercury is found in all mercury mines. It occurs in small drops, attached to the body of the ore, to the gangue, or dead minerals of the vein, and to the rock. The most important quicksilver ore is the

Sulphuret of mercury—cinnabar.—This mineral resembles, in colour, oxide of iron, with which it is sometimes confounded. Its redness, however, is mingled with a yellowish hue, like that of minium, by which peculiarity it is distinguished. It is also easily distinguished from other minerals by its volatile character. It evaporates entirely, when thrown on red-hot iron, leaving no residuum, and emitting a strong smell of sulphur. The powder of cinnabar, when rubbed on gold or copper, whitens these metals, as if plated with silver. Cinnabar is found in California, and is said to occur in heavy masses. The ore is of a beautiful appearance, pure, and compact. It contains 84 per cent. of metal, and 14 per cent. of sulphur. High wages, and the want of roads in that new State of the Union, have prevented till recently the working of these quicksilver mines; but if the population of California continues to increase as fast as it has done since its annexation to the Union, there is little doubt of the cinnabar mines soon being worked.

Bituminous sulphuret of mercury.—This is a variety of cinnabar, of a more or less grey, brown, earthy colour and appearance. It generally accompanies the pure qualities, and is mainly distinguished from them by its colour. On heating this quicksilver ore, it emits a very disagreeable smell, and leaves a residue of earthy matter.

There are other ores of mercury, but they are of little importance. They may all be distinguished by their entire volatility, and their capacity of coating gold white when rubbed upon it.

The geological position of the quicksilver ores is in the older rocks of the secondary formation, or the later series of the transition rocks. We find it therefore in the New England States, along the Lakes, and in the gold region of the Southern States, but in such small quantities that the mining of it cannot be carried on to advantage. Quicksilver ores may also be found in an earlier series of rocks than the bituminous coal; but in that case there is some metamorphic or volcanic rock in its vicinity, which appears to have been the means of depositing it where it is found. These ores are very volatile, and any volcanic eruptions will bring them from the interior of the earth, to condense on some convenient spot, colder than the place of their origin. Quicksilver ores are not always found in regular veins. We find these ores in grains, disseminated through the masses of rock, like gold, platinum, and other metals. One of the quicksilver ores of Spain is a black slate impregnated with metallic quicksilver.

CHROMIUM.

Chromic iron — Chrome ore; chromate of iron.— Chromium is not found as a native mineral. The only ore of practical use of this kind of metal is the chromic iron, of which the richest deposits in the world are in this country. The bare hills near Baltimore supply almost the whole world with chrome ore. The colour of this ore is a brownish-black, like a hard brown iron-stone. It resembles black manganese, but is harder, and of almost metallic lustre. Its powder is brown; and in the mass it is brittle. Besides Baltimore, the ore is found in Pennsylvania, New Jersey, and most of the New England States, particularly in Vermont, and near Chester, Massachusetts, where it is deposited in considerable masses. The best qualities of this ore contain 60 per cent. of oxide of chromium; those from Baltimore average 40 per cent: the remainder is iron, clay, and siliceous matter. Chrome ore is only found in serpentine rocks, forming veins, masses, and small pockets. It is not found in granite, nor in the coal formation and the rocks of a later period.

MANGANESE.

There are only two principal ores of this metal which are of practical value; the others are merely objects of science. Manganese is never smelted by itself, but is a constant companion of iron. There is scarcely an iron ore which does not contain more or less of this metal; and there is little iron, particularly pig-iron, free from it. The most abundant ore is the

Black manganese—peroxide of manganese.—This is found in all the States along the Atlantic coast, near the Lakes, and west of the Mississippi river. Black manganese consists of 63 per cent. of metal, and 36 per cent. of oxygen. It is a very dark-brown mineral, and generally has a velvety appearance where it has been exposed to the atmosphere. In the fresh fracture it is close, compact, and of a vitreous lustre.

The other variety of manganese has a more bluish cast, which is caused by admixtures foreign to the ore. It is much harder than the first. Both varieties are used chiefly in glass-works, to clear or bleach the glass. For this purpose it is necessary that it should be very pure, and especially free from iron, with which it is nearly always adulterated.

There is still another variety of manganese ore, which is found in a crystalline form. It possesses a metallic lustre, is of a brownish-grey colour, and affords a black powder. This is the common manganese, and is more used than the above kinds. It is particularly employed for making chloride of lime, and also in the glass-works.

Vermont formerly furnished the best article of manganese in market; but inferior qualities are now sent from that State. Virginia, Pennsylvania, and other States also furnish manganese. It is found in every geological formation, in or near the oldest rock, in volcanic regions, and in alluvial gravel.

ZINC.

Native zinc is never found; it has so much affinity for other matter, particularly oxygen, that it cannot exist very long in its pure state. The following are the principal ores of this metal, arranged according to their quantity in this country:—

Zinc-blende, or simply blende.—This is a sulphuret of zinc, and is composed of 68 per cent. of zinc, and 32 per cent. of sulphur. This ore is always found crystallized; and in most cases the masses of it are mere accumulations of crystals. Its colour is generally a bright or yellowish-brown; but it is occasionally found to be black, red, green, or yellow. It is transparent, or at least admits of the passage of light, if in thin splinters. The lustre of this ore is brilliant, and more decidedly adamantine than any other ore. It is found in heavy veins and masses in the gold region of the Southern States, where it forms the principal silver ore. It also contains a considerable amount of gold. It is here associated with galena, iron and copper pyrites, tin, heavy spar, black manganese, and manganese-spar. The bulk of this ore is blende, which in most cases constitutes at least 50 per cent. It is chiefly worked for its silver and gold. All the other metals, and particularly zinc, which could be manufactured cheaply from it, are lost, being either washed away or converted into cinders.

Blende occurs chiefly in the rocks of the transition series, and is abundantly found in the United States. The operation of smelting this ore, and extracting from it the gold, silver, and lead, is extremely troublesome.

Red zinc ore—zinc ore—is extensively deposited in New Jersey and Pennsylvania, and is used for the manufacture of brass. This ore is a compound of oxide of zinc, manganese, and oxide of iron. Its colour is a brick-red, with a yellowish tinge, like cinnabar. Its texture is granular and massive. This ore, like the blende, belongs to the transition series of rocks.

Calamine—silicate of zinc—is very widely distributed, and is found in heavy beds in eastern Pennsylvania. There are two kinds of this ore, the silicate and the carbonate: the latter species is that alluded to as found in Pennsylvania. This ore appears in kidney-shaped masses, and in concretions like iron pipe-ore, honey-combed. It is of a dirty yellow or stone colour. If pure, it consists of about one-half oxide of zinc; the other half is composed of carbonic acid, silex, iron, water, and other admixtures. It is found in heavy veins, and may be looked for in all limestone rock, from the most recent to the oldest formations.

Little or no use is made of zinc ores in the United

States. The low price of the European metal prevents the erection of smelt-works on a scale commensurate with the amount of ore at our disposal. Specimens of American zinc have been exhibited, superior in quality to any from the old world.

TIN.

There are but two ores of tin known, which are of any practical use. One of these is tin-stone, or peroxide of tin; the other is sulphuret of tin, or tin pyrites. No tin is manufactured in the United States at present; still, there are indications of the ore, and, at some future day, it may be found advantageous to smelt it.

Tin-stone is found in the New England States, and is said to have been recently discovered in Missouri. This, however, is doubtful. Tin-stone occurs chiefly in granite, in heavy masses or lodes, mixed with conglomerates of various rocks. It is also found in alluvial gravel, as the result of the decomposition of the above rock, and is then called stream-tin. Tin-stone is of a variety of colours, white, grey, yellow, red, brown, and black. Its most striking feature is its weight. Its specific gravity is equal to that of galena, from which it is easily distinguished. This ore is very hard, of a brilliant lustre in the fresh fracture, and is frequently found in detached double crystals. By striking with this stone upon steel, fire can be produced.

Tin pyrites are not very abundant, and cannot be considered an ore of tin; their presence in the silver ores of the southern gold region is so limited, as to render the extraction of the metal unprofitable. This ore is of a grey or yellowish colour, heavy, crystallized, and of a metallic lustre. It is always found to be adulterated with foreign matter, as iron, copper, lead, and other ores, which, in the above-mentioned case, predominate so much as to make the smelting of tin from these ores impracticable.

There are a number of useful minerals, or ores, of which it is beyond our limits to speak. These are, properly, subjects for works of a more scientific character than this essay pretends to. The ores of Nickel, Cobalt, Antimony, Arsenic, Bismuth, and a variety of others, are not distinguishable without a chemical analysis, which it is not our intention to give in this work. Such minerals must be left to the analytical chemist to decide their composition. Among the useful minerals, however, there is a class of which we shall speak, namely, those which do not constitute ores of metals. For the sake of convenience, we will follow an alphabetical order, in describing and enumerating these minerals.

ALUM-SLATE,

Also called alum-schist. This is the material most extensively used for making alum. This is a black or brown coal-slate. It is also a clay-slate, containing so much bituminous matter as to render it combustible if piled up in large masses and kindled. The alum-slate ought to contain clay, which is indicated by its close and fine-grained texture; carbon, to render it combustible; and, what is most essential in its composition, pyrites, to afford sulphur for the formation of sulphuric acid. The dark colour of a slate indicates the presence of bitumen or carbon; and yellow streaks or dots, of a brass colour and lustre, show the presence of pyrites. Beds of alum-slate are found almost everywhere in the United States, particularly in the western coal fields, where it abounds. On and near the Ohio river, alum is manufactured to a large extent, from alum-slate of the coal region.

ALUM-STONE.

This is a rocky, volcanic mass, used in Italy for the manufacture of Roman alum. This stone, the result of volcanic action, is impregnated with alum in its finished state, but in a basic form; that is, it contains too much earthy matter. The mere roasting of this stone transforms the insoluble basic alum into soluble common alum. I allude to this rock, not knowing whether it is found in the United States, because there is a similar basic alum found in the bituminous coal fields of the Western States. The waters drained from old coal pits, in these regions, are often surcharged with alum. This is a basic alum, in which the potash is replaced by oxide of iron. Such alum appears also as a white exuberance or powder, or in small crystals, in the coal mines and the cavities of the coal rock.

AMBER.

This mineral is not found to any extent in the United States; but it will not be without interest to draw the attention to it. Amber is a solid, resinous mineral, of various shades of colour, but mostly of a brownish yellow; it is generally transparent, but there are also opaque specimens. Its most characteristic feature is its fine aromatic odour, which it emits when thrown on hot coals. It is found in kidney-shaped concretions in the sands along the sea-coast, or in the vicinity of layers of brown coal; it is often found imbedded in the latter substance. Amber is, properly speaking, a rosin, and is the most valuable of this class.

ASPHALTUM,

Bitumen, Petroleum, Naphtha, Rock-tar, is a fluid mineral substance, found in some places in the Union. It exudes from the earth along with water in springs, near rock salt or brine springs. It burns with a smoky flame. All mineral resins are of a similar composition; they are found both in a fluid and a solid state, and are chiefly valuable as fuel. Distilled naphtha is used in the arts.

ASBESTOS,

Mountain wood, is a siliceous mineral found on the eastern side of the Alleghenies; it is characterized by its resistance to the influence of heat. It is a fibrous mineral, often appears as if composed of fibres, and is easily split in the direction of its fibres. It is very elastic and malleable, has a silky lustre, and feels as soft as silk. Asbestos is of various colours, white, green, and brown; and is mostly found imbedded in serpentine rock and mica-slate. The inferior qualities are used as fireproof material in smelting-furnaces, and called soapstone. Lamp-wicks and the linings of fireproof safes are also made of this material.

CHALK.

White chalk is a very friable limestone, soft and dull. It is distinguished from clay by non-adhesion to the tongue: it will not receive an impression, does not possess the plastic and tenacious character of clay, and cannot be polished. When chalk is purified by washing and grinding, it is called Spanish white.

The United States afford but little, if any, chalk; what we have is chiefly imported from Europe.

Black chalk—drawing slate—is a slate containing carbon, so soft as to make a grey or black mark on any white substance, or when rubbed on paper. The deposits of alum-slate afford samples of black chalk.

Red chalk is a peroxide of iron, very much contaminated with clay or siliceous matter. It contains generally but little iron, and cannot properly be considered an iron ore.

French chalk.—This is a fine quality of soapstone, found in great abundance in the Atlantic States. It is a soft, friable, magnesian slate, appearing in veins, imbedded in mica-slate and talc-slate.

CLAY,

Argillaceous earth—Potters' earth.—There is a great variety of this class of minerals. All clays are distinguished from other earths by containing a certain amount of alumina, which imparts to clay its remarkable adhesiveness. The common characteristic of clay is, its ready diffusibility in water, in which, although insoluble, it remains suspended longer than any other mineral. All clays are plastic, that is, may be formed into various shapes; and the figures thus made will retain the shape given them, will harden by exposure to atmospheric air, and can be solidified to such a degree by heat as to yield fire on being struck by a steel. It shrinks while drying, and the more alumina it contains the greater will be the shrinkage. Clays impregnated with vegetable matter, that is, all plastic clays, emit a disagreeable odour—an argillaceous smell—when breathed upon. Plastic clay, when dried in the air, or not burned too hard, adheres to the tongue, in consequence of its affinity for moisture. Clay is the residuum of decomposed rocks, its great adhesiveness causing it to be found in beds of considerable extent, and with a well-defined line of separation from other matter. Its affinity for vegetable matter, bitumen, and some metallic oxides, particularly oxide of iron, and for acids and salts of various kinds, renders it a useful material in the arts. This affinity for other matter causes, sometimes, great variations in its appearance. The United States abound in all the different varieties of useful clays, from the best kinds of fire-clay to common loam.

Loam—common clay—is the most generally diffused; it contains but little alumina, and is chiefly composed of silica or sand, with lime, iron, magnesia, mica, and a variety of other substances. Loam is more or less adhesive, and may be worked into a tenacious paste; but it is generally more brittle when dried than the pure kinds of plastic clay. To this class belong all the blue, yellow, and red clays. Loam is fusible at a high heat, and melts into a black, tough cinder. It fuses most readily when united with iron, lime, or other metallic oxides. Some clay contains common salt, saltpetre, or phosphates and sulphates, which render it more fusible. If clay contains any soluble salt, it may be discovered by drying it, when the salt will crystallize, and be visible on its surface, in the form of an efflorescence, where the tongue can easily discover the kind of salt it contains. Loam is used in the manufacture of bricks, tiles, and the coarser kinds of potters' ware. The distribution of this clay is so extensive, as to preclude the possibility of our entering into an enumeration of its localities. Some varieties of loam are much better adapted than others to certain manufacturing purposes. The Atlantic coast abounds in fine loam, suitable for making bricks; and heavy deposits of London clay are not uncommon. In the coal region of the West there are fine loams, but they are rather fusible. The red marl strata of the coal series—a red clay—is very plastic, but is easily melted. The best clays are generally found on hill-sides, or on their summits.

Potters' clay—plastic clay.—This species is more compact, more tenacious, and smoother than loam. It may be polished by rubbing it with the finger, is exceedingly ductile, and takes the impression of the hand perfectly. Mixed with water it forms a somewhat transparent paste. This clay is always white or grey, does not alter its colour when exposed to a red heat, is infusible in the greatest heat of a porcelain oven, and can be made exceedingly hard by the action of fire. When produced by the decomposition of feldspar, and when it has a gritty feel, it is called kaolin, which is the principal material used in the manufacture of porcelain.

In the selection of clay for fine ware, some judgment is required. If the clay assumes any colour,—even the faintest shade of a colour,—on being exposed to a white heat, it is not of the right description; the ware manufactured from it will be brittle, or will not endure a heat sufficient to harden it, without an alteration of the form. Good, plastic clay, is extensively distributed throughout the country, particularly on the Atlantic coast, and in the Western States. Its geological position is always lower than the loam, or below the London clay, and above the rocks of all ages. It is thought to be the product of decomposed slate-rock, or of primitive rock.

Kaolin—porcelain clay—china clay.—This is formed by the decomposition of feldspar and granitic rocks. It feels gritty, is friable and meagre, and not so diffusible in water as plastic clay. Kaolin is infusible in the most intense heat of any fire, retains its white colour, and does not shrink so much as the plastic clay. The chemical composition of both of these clays is very much the same when pure, each consisting of about one-half alumina and one-half silica. Kaolin is very common on the eastern slope of the Alleghenies, but is seldom found on the western side. The New England States furnish kaolin, New Jersey abounds in fine specimens, and all the Southern States produce it of a superior quality, North Carolina appearing to be especially favoured. Kaolin is the best material known for the manufacture of porcelain; it is more expensive to work than plastic clay, but the ware is stronger and whiter than if made from any other material.

Slate-clay — fire-clay. — This material is found in veins in the coal regions, generally below the coal formation; but it is also found lying between, or above the seams of coal. The veins in the western coal fields are frequently interspersed with nodules of argillaceous iron ore. This clay breaks in the form of short slate; it falls into small square grains, and is converted into plastic clay by the action of the atmosphere: this may be seen where the veins outcrop. The colour of this clay is grey, or dirty yellow; it is soft, and easily broken, adheres slightly to the tongue, and dissolves slowly in water, being converted into a clay paste. This kind of clay is ground, and made into fire-bricks, for which it is admirably suited; it is extensively used, particularly in the Western States. A fine quality of this kind of slate-clay has been found, and is now worked, in the Maryland coal basin, near Frostburg, where a superior article of fire-bricks is made from it. It is much used along the Allegheny and Ohio rivers; and, at the Conemaugh river, near Blairsville, Pennsylvania, there is a heavy vein of the first quality, from which fire-bricks are also manufactured. Veins of this clay often contain lime in admixture, which injures its fire-proof quality. Lime, if mixed with it, is found at the top and at the bottom of the vein. All of this kind of clay contains a little iron, which offers no impediment in the making of fire-brick, but renders it useless for the manufacture of fine porcelain.

Pure clay is infusible in any heat; but if it is mixed with a large proportion of iron, sand, or lime, it then becomes fusible. Common loam is of this composition, and is fusible. If loam or clay is not sufficiently refractory, it may be made so by mixing pure sand with it; and some kinds of loam will bear the addition of an equal volume of sand, before they will make a good common brick. Different descriptions of clay are coloured blue, grey, or yellow; all such clays turn red or brown in the fire. In many clays the colouring matter is not apparent, until they are exposed to a heat, almost sufficient to melt them; other clays are red, when fresh, but lose their colour on being heated to a cherry-red heat, assume a brighter colour on the application of a more intense heat, and are at last converted into a black slag. To make bricks of such fusible clay, requires the addition of a large proportion of sand. The fattest kinds of clay are the most liable to crack by shrinkage, which is a serious obstacle to their use in the manufacture of ordinary pottery, and only remediable by the addition of sand. Hard-burnt stone-ware, powdered, and incorporated with the clay, is, however, better suited to this purpose than sand.

COAL.

All mineral substances containing sufficient carbon to supply their own fuel, and support combustion, may be called coal—mineral coal. There are vast beds of slate and limestone, containing a large amount of carbon, which cannot properly be classed as coal, because they do not perpetuate combustion.

Charcoal.—Mineral charcoal is frequently found in bituminous coal, in the form of thin seams and plates; but the quantity is so small as to render its separation from other coal impracticable.

Anthracite—hard stone-coal—is, with the exception of charcoal, the purest mineral carbon of which we have any knowledge; and it forms the bulk of the fuel used in the Atlantic cities. It is found in immense bodies in Eastern Pennsylvania; also in the New England States, and in North Carolina. Anthracite coal forms heavy veins and masses in the metamorphic rocks of the eastern slope of the Alleghenies, but is seldom found on the western side of that mountain chain. Hard coal is a very black, hard substance, of great lustre; it breaks in irregular fragments, and is not affected by the atmospheric air. The chemical composition of this coal is, almost entirely pure carbon, a little hydrogen, and a small per centage of ashes. When a sufficient quantity of this coal is ignited, it creates an intense heat; but, in small quantities, it does not burn well, and requires a strong draft to support combustion in a small space.

This coal is also more difficult to kindle than any other, requiring the use of wood, or wood-charcoal, to ignite it. This fuel is well adapted to smelting operations, for which purpose it is extensively employed. The geological position of anthracite coal is said to be in the transition rocks; but, in this country, we cannot place it in that series.

Bituminous coal—soft stone coal.—This name is applied to a mineral coal, which has never been properly defined. A natural and well-marked distinction of this coal is, its property of coking; that is, if exposed to a red heat, it blazes, swells, and finally bakes together, forming a spongy mass, called coke. All the coal which is black, and makes a black powder, but does not coke, is anthracite; and all the coal which is soft, and makes a brown powder, but does not coke, is denominated brown coal. Bituminous coal is black, makes a black powder, has a bright, resinous lustre, and is liable to form slack when exposed to the air. It is distinguished from anthracite by possessing a more slaty structure. Some of this coal breaks into beautiful cubical pieces, of which the Pittsburgh vein shows many fine samples. Most of this coal is inclined to break in that way, but the cubes are often very small, and form mere slack. Bituminous coal burns easily, with a bright, vivid flame, similar to that of pine wood. Anthracite usually burns without any flame; but, sometimes, with a faint, blue, scarcely-visible flame. Bituminous coal is easily kindled, and, in small quantities, supports combustion with facility. The chemical composition of this coal differs from anthracite, only in the larger amount of hydrogen it contains: this, in combination with carbon, forms bitumen, which can be extracted by distillation in iron retorts; and from this circumstance is derived its name, "bituminous coal."

In the Mississippi valley, that is, in the region where this river and its tributaries flow, the amount of bituminous coal buried beneath the surface of the earth is so large, that no parallel can be drawn between the amount of coal in that district, and what is contained in all the other parts of the world. The quality of the coal in this large basin is, as may be expected, very different. West of the Alleghenies, the Pittsburgh coal is considered to be the best, that is, the coal from a vein called "the Pittsburgh vein," which is, notwithstanding its local appellation, of vast extent. This vein may be traced to a distance of from one hundred to one hundred and fifty miles from Pittsburgh, and furnishes at every point where it has been opened, the same kind of coal. There are indications that the thick Frostburg vein, in Maryland, forms a portion of the Pittsburgh vein; which fact, if corroborated, would extend the Western coal field through the Allegheny mountains. The coal from the Pittsburgh vein, of which the coal from the Youghiogeny river, a branch of the Monongahela, may be considered the finest, is a beautiful jet-black coal, almost free from sulphur, of a high lustre, frequently displaying the colours of the rainbow. This coal breaks into cubes, of from two to six and eight inches. It is not very liable to form slack on being exposed to the air. It makes a beautiful, strong, and pure coke, and may be considered the best coal for making gas. This vein does not furnish an equally-beautiful coal throughout its extent, still, it produces in every part a superior coal. Within one hundred miles of Pittsburgh, along the Monongahela river, this vein produces superior coal; below Pittsburgh, on the Ohio river, and above Pittsburgh, on the Allegheny river, this coal is not so good as at the first-mentioned place. Pittsburgh may be considered very near the centre of the large western coal field; and here the coal, particularly that of the thick vein, is superior to any coal, no matter whence it comes. The same vein furnishes a fine coal along the borders of the coal field, where the vein is generally from ten to fourteen feet thick—near Pittsburgh it is but seven feet. This coal cannot be sold where Pittsburgh coal can be had at an advanced price. The geological position of this coal forms a particular era, a separate period in the formation of rocks. It is not found in the old rocks — in granite and its associates — nor do we find it in transition or metamorphic rocks; yet, in Virginia, a bituminous coal is found imbedded upon granite, and surrounded by the younger transition rocks. This coal cannot be considered a true bituminous coal, but forms a link between this and the next following — brown coal.

This Virginia coal burns with a vivid flame, and forms a sort of friable coke; still, we cannot consider it as a true bituminous coal, of the secondary formation. It is in vain to look for bituminous coal in the transition and metamorphic rocks on the eastern slope of the Alleghenies, or along the Lakes, and among the Rocky mountains. Lignite and anthracite coal may be, and is found in these regions, but no true bituminous coal. For domestic use, and also under steam-boilers, it does not make much difference whether we burn lignite or true bituminous coal; but it makes a great difference in metallurgical operations, because lignites, even the best kinds, are generally very sulphurous, which is considered disadvantageous in smelting operations. Metallurgical works must be carried on in the anthracite districts, or in the coal formations of the secondary strata, if expected to be prosperous.

Brown coal—lignite.—This is a kind of coal more generally distributed than known. This coal is found among or above the rocks of all ages. We find it resting on granite in Virginia, on the James river, imbedded in tertiary rocks along the Atlantic coast, in Michigan and Missouri, in Oregon, on the shores of the Pacific, and in California. We find this coal jet black, and also of a brown colour; we find it as hard as anthracite, and so soft as to crumble into fine dust on being touched. It is never found in extensive layers, like the coal of the secondary strata, and the deposits mostly form thin veins or elliptical masses.

This coal is characterized by its making a brown powder, which may be more or less dark, but always shows its brown colour. Some of this coal forms coke, but the coke is weak, friable, and, on account of its impurities, not qualified for smelting iron and other metals. The greater bulk of this coal does not coke at all. It always contains more impurities than coal of the older rocks, makes more ashes, and is very sulphurous, which causes operate in the formation of clinkers in the fire-grates, causing trouble to the firemen.

If brown coal is very impure, slaty, and does not contain sufficient carbon to constitute fuel, it can be used for the manufacture of alum, either by burning or roasting in large piles, and may be considered the best material for that purpose. We may expect to find brown coal below the sand of the beach, as well as in the granitic and volcanic mountain regions. It is always accompanied by fossil remains of plants and animals. Coal which is too brittle to bear transportation, and diminishes the draft of a fire by falling into dust, may be moulded like fire-bricks, dried, and then used as fuel.

There is some uncertainty in the public mind, as to the real value of the various kinds of coal; and a brief discussion of the subject cannot but prove interesting to all. With reference to all the varieties of mineral coal, particularly hard coal, and true bituminous coal, we may say, "the true value is represented by the combustible matter." As such coal does not contain any water, the amount of ashes, or the unburnt residue, is the only deduction from its heating effect. If there is any difference between hard coal and soft coal, of equal purity, it is in favour of hard coal; for, the less hydrogen fuel contains, the greater the amount of heat liberated from the same weight. There are, however, differences of a practical nature, which considerably modify this theoretical fact. A fire-grate, or furnace of any kind, which burns soft coal to advantage, is found to have a contrary effect upon hard coal; and the grate in which hard coal can be profitably consumed, will not answer so well for soft coal. Soft coal burns more rapidly, and distributes more heat in the same time, over the same space; but it also consumes faster than hard coal would, under similar circumstances. The absolute amount of heat evolved, is greater in soft than in hard coal, but this heat cannot be used to so much advantage as that produced from hard coal.

DIAMONDS.

Diamonds—pure carbon—are rarely found in the United States. A few have been picked up in the alluvial gold-washings of North Carolina, as well as in the other States composing the gold region. In Georgia, diamonds are more frequently found; and, but a short time since, a diamond valued at one hundred and fifty dollars was discovered in this State.

California, also, produces some very fine specimens. As there appears to be no correct information as to the localities in which diamonds may be sought for advantageously, it may be stated, as a general rule, that diamonds cannot be found in the coal regions. Where gold, and particularly coarse gold, is found in alluvial soil, there is a probability of the presence of diamonds. The geological formations of the gold regions, speaking generally, possess strong indications of their presence; but they are always found at a much greater distance from the surface of a gravel-bed than gold. The matrix of diamonds appears to be that gravel of the old formations, known by the name of "pudding-stone," in which quartz pebbles are cemented together by the oxide of iron. If it is necessary to wash gravel, in order to discover the diamonds contained therein, it should be done in the bright sunlight, where they may be detected by their sparkle.

FULLERS' EARTH,

Is a clay of a particular composition, containing principally silex, some alumina, oxide of iron, magnesia, lime, and other substances. Any fine clay which dissolves, at a cherry-red heat, into a black or brown slag, may be called fullers' earth. This material abounds in every part of the United States, the green strata of clay, containing iron, of the western coal fields, being not the least useful.

GARNETS.

Garnets are very extensively distributed, particularly in the gold regions. It is a very hard mineral, of a dark red or brown colour, and is found in crystals of the form of a dodecahedron, or nearly round, and from one-twelfth of an inch to one or more inches in diameter, disseminated in rocks. The finer qualities are employed in jewellery, the less brilliant specimens are pounded, and used as emery. There is an abundance of garnets in the Southern States, and, by converting them into emery, a prosperous business in that article might be established.

GRAPHITE,

Plumbago—black lead—is a mineral of a lead or iron-grey colour, and metallic lustre; it is soft, and easily rubbed into powder, or cut with a knife. Its powder stains the fingers with a lead-grey hue, and feels fatty. It is a crystallized carbon, and, when freed from accidental foreign matter, is a pure carbon. The United States afford an abundant supply of this mineral; it is found in almost all the Atlantic States, and also in some of the Western States. The geological position of this substance is in the primitive rocks, where it is found in veins and pockets. In Virginia, North Carolina, and other States, it is found imbedded in mica-slate or talc-slate. Graphite is used for counteracting friction, making lead-pencils, blackening and polishing cast-iron stoves. The most important use to which it is applied is, the manufacture of crucibles, for the smelting of brass, iron, and other metals.

GRINDSTONES.

Any fine-grained sandstone, containing a large portion of alumina, will make a good grindstone. The sandstones of the transition series are usually harder and more durable than those of the younger rocks. In selecting sandstones for this purpose, it is desirable that they should be entirely free from fissures or stratification, and possess such an intimate adhesion of their particles, as to resist abrasion. Fine-grained sandstones having great strength and uniformity in their mass, are preferable for this purpose to the coarse-grained material.

Whetslate is a hard and fine variety of grindstone, the most superior of which are called oil-stones. These slates are white, grey, and all varieties of colour, the greenish-grey being generally the best. Its fracture being slaty or splintery, it breaks into tabular or slaty fragments, is translucent on the edges, and exceedingly hard. Material of this kind is very extensively distributed throughout the United States; but the best quality is found in Virginia, and in the range of the gold region. The stone called turkey-hone, so justly celebrated for imparting a good edge to fine instruments, is a variety of dolomite, and receives its excellent qualities from saturation in oil. To form good oil-stones from a rock, the rock should be broken, not sawed; for, in all such rocks, there are veins of soft and hard material, or even fissures. If the rock is sawed, all these imperfections may be found in the small hand-stone; but if it is broken into small fragments, it will naturally part at the weakest points.

Emery.—The hard varieties of dolomite—whetstone—form an excellent emery, which is used for grinding and polishing hard steel and glass. The hardest emery, but one not much used, owing to its expensive character, is pounded korund, a material similar to the dolomites of the gold region. There is an abundance of material in the Southern States, for the manufacture of emery, requiring nothing but the necessary enterprise and energy, to enable us to supply the home-market, and stop the importation from Europe.

Flint is almost entirely pure quartz; it is excessively hard, but by a skilful operation it is cut into sharp-edged fragments, which, being struck smartly on steel, will produce fire. The Indians made use of it for their arrow and spear points, and also for their hatchets, great numbers of which are found in the soil, in some parts of the Union. Flint is found in nodules, imbedded in chalk; but the United States not producing much chalk, we cannot determine whence the Indians obtained the flint which they used. The fine powder of flint is used for grinding purposes, but it is as expensive as it is useful. This powder is an excellent material to mix with clay, for the manufacture of potters' ware.

HEAVY SPAR.

Sulphate of barytes—is found in many States of the Union. The colour of this mineral is generally white or flesh-colour; it is found in a crystallized form, and possesses a remarkably great specific gravity. Its white variety is commonly used in the adulteration of white lead.

LIMESTONE.

Lime is a generally useful material, and is, fortunately, very abundantly deposited throughout the Union. The purest limestone is a calcareous spar, and is found in white crystals, in the crevices of limestone rocks, in veins of ore, or in the crevices of any description of rocks, filling the spaces, wholly or in part. Another species of pure limestone are stalactites or calcsinter; these are found in the caves of limestone rocks, or are concretions formed by the sediment of spring water. Stalactites are not of much practical use; the pure or clear specimens, being faintly transparent, are cut and turned, by the help of a lathe, into small ornaments or busts, which is almost the only use that can be made of them. The foregoing and the Carrara marble, are the only specimens of pure limestone, or carbonate of lime; the following are, more or less, impregnated with foreign matter.

Compact limestone.—This is a class of limestone, of which there are many varieties. Its grain is more or less coarse, it does not take a polish, like pure limestone, nor will it break into blocks sufficiently large and sound to be used for ornamental purposes, such as slabs or statues. The colours of this stone are various, and range from a yellowish white to a jet black. This limestone is extensively distributed over the Western States, and is also found in thick and extensive veins, imbedded in the coal strata. In the States east of the Alleghenies, it is found in extensive veins and deposits. In the limestone rocks of Missouri, Illinois, Virginia, and other States, galena is found in exhaustless quantities. This limestone is a pure carbonate of lime, containing matter foreign to lime, chiefly silica, alumina, magnesia, iron, manganese, carbon, and other substances, in variable quantities; so that it appears to be a limestone embodying foreign matter or siliceous rock, containing lime. The limestone of this class is the most used in the manufacture of lime, the quality of which depends in a great measure on the purity of the stone.

Magnesian limestone.—Some varieties, as those in and below the anthracite region, and in some places on the Ohio river, contain a large amount of magnesia, and are then called magnesian limestone.

Others contain a large quantity of oxide of iron and silica, and, consequently, will not slack when burned and moistened. When burned and ground, it forms an excellent water-proof cement. This limestone is found in veins of from one to five feet thick, in the bituminous coal fields, forming round masses of different sizes. In Kentucky, this limestone rock contains extensive subterranean cavities, and is then denominated cavern limestone. In other places, especially in the State of New York, it has a stratified appearance.

Lithographic stone.—The fine-grained limestone of this kind, which breaks into thin slabs, is termed lithographic stone, because it is used for multiplying impressions of drawings, made or engraved upon it. Some specimens of it have been discovered in Alabama, but it does not appear to be of a sufficiently good quality to substitute for the German article.

In addition to the limestone belonging to the foregoing class, we have the marly limestone, bituminous limestone, and a variety of others, discoverable in the tertiary rocks. These are, however, owing to their impurity, seldom used in the manufacture of lime. The limestones of this formation generally furnish a good material for making Roman, or water-proof cement.

Shelly limestone is composed principally of shells; it is of various colours, but is most commonly either grey, white, yellow, or brown. This limestone is never pure, always containing silex, alumina, iron, and other substances. It is excellent when used for fertilizing the soil, and is then termed calcareous marl. It is found in heavy veins in the coal formations, and in tertiary rocks along the sea-coast.

Gypsum—plaster of Paris—is not properly a limestone, but its base is lime, and it has, therefore, a right to be mentioned in this place. Gypsum is composed of lime and sulphuric acid, whilst limestone contains lime and carbonic acid. It is found in abundance in the Atlantic States, and also in some of the Western States; it very much resembles limestone in appearance, but is distinguished from it by breaking in irregular blocks, the fracture having a glistening appearance. It is frequently impregnated with mica, talc, oxide of iron, manganese, and other matter, which may be distinctly seen, disseminated through the rock, in grains or spangles. Gypsum is found among and near the rocks of all ages, and is a constant accompaniment of rock-salt. When gypsum is exposed to a gentle red heat, it loses its water of crystallization, falls into a fine white powder, and becomes what we term plaster of Paris; and this, when moistened with water, quickly hardens into a white, solid mass. Gypsum is an active agent in fertilizing the soil; and burned gypsum is extensively used for plastering and ornamenting the interiors of houses.

Burned lime, or *quicklime*, is employed in the arts for an infinite variety of objects. It is used for fertilizing land, making mortar and cements, clarifying the juice of the sugar-cane, purifying coal-gas, bleaching cotton, making soap, and for many other purposes.

Marble is a species of limestone, which can be quarried in such large and sound blocks, that it may be cut into slabs, or used for statues. There is but one distinction between common limestone and marble, the latter is sound and compact, a quality which the first does not possess. All limestones, furnishing large blocks, without fissures and stratification, may be termed marble. The distinguishing quality of marble is a fine grain, adapted to receiving and retaining a fine polish. The colour or colours of marble should be distinctly expressed, and not have a dull or earthy appearance; white marble should have a clear, pure colour, and not one of a dirty-grey character: the same rule is applicable to other colours. If the marble should be variegated, its value will be determined in a similar manner: all the colours of a variegated marble must be distinctly expressed, to make it worth the highest price. The United States afford an abundant supply of marble, but comparatively little of the fine-grained statuary marble. Most of our marbles are of the foliated kind, or a conglomeration of large crystals, which makes them useful for architectural purposes, but excludes their use in the finer works of art. The greater proportion of our marbles belong to the transition and coal formations, and cannot be placed in the first class. Some fine saccharine marble has been discovered in the New England States, but we are not aware that it is much used. A fine variety of variegated pudding-stone, or breccia marble, is found in Maryland. In opening a marble quarry, the first consideration should be its locality, and the means by which heavy blocks may be removed from the quarry. High prices are paid for fine-grained, pure white, saccharine marble; and the statuary marble imported from Carrara, in Italy, is frequently sold as high as fifteen dollars per cubic foot.

MARL,

Is becoming a very important agent in fertilizing the soil. The heavy deposits of marl along the Atlantic coast, among which those of New Jersey are the most extensive, have been mainly the cause of its application to agricultural purposes: its use has been followed by unexpected success. The green-sand marl of New Jersey is a union of silex, potash, clay, and a little iron; it is a green, coarse-grained sand, turning dark green, or bluish green, on exposure to the air. This marl is found in a stratified state, imbedded in the tertiary rocks. The New Jersey marl is distinguished from nearly all the other kinds of marl, by the large amount of potash which it contains — being about ten per cent. of its weight — which is not the case with the others, and pre-eminently qualifies it for the assigned purpose. Marl is generally composed of siliceous matter, lime, magnesia, alumina, and iron, in variable proportions. In every case, marl can be successfully used in the improvement of a sandy soil. The United States abounds in marl of all descriptions, and when its use is more generally understood, it will become a useful agent in advancing the agricultural interests. The bituminous coal region of the Western States, affords an inexhaustible supply of aluminous as well as limestone marl. Heavy beds of marl are found in Virginia, and the Southern States.

OCHRE.

Yellow ochre is an impure hydrated oxide of iron, in which silex and alumina form the bulk, and iron, the colouring matter. This mineral is found in layers or veins, and is prepared for use by grinding it under mill-stones, working edgewise. Yellow ochre, when burnt or calcined, forms red ochre or Spanish brown. Native red ochre is red chalk, or reddle.

SALT.

Common salt—sea-salt—rock-salt.—This mineral is found, in a solid form, in rock-salt, and, in solution, in sea-water and in brine-springs. Rock-salt is found in North Carolina; but the principal source from which the Atlantic States obtain their supply of salt is, the evaporation of sea-water. In the Western States, salt is made by the evaporation of the brine found in the lower strata of the coal formation. Artesian wells are sunk in the rocks, to a depth of from one hundred to twelve hundred feet, the brine is raised by means of pumps, and the salt obtained by evaporating the brine in iron pans. Natural brine-springs are found in the State of New York. There is an extensive lake of concentrated brine in the Territory of Utah.

SAND.

Sand, on account of its extensive use, is a very valuable mineral. It is generally a hard, granular substance, consisting principally of quartz; it is found in the form of strata, and in loose, alluvial deposits, of immense extent. All siliceous sands appear to have been originally in the form of crystals, which have been rounded and worn off by friction, occasioned by the rubbing of one grain against another, while suspended in water. In coarse sand, or gravel, we frequently find well-preserved pebbles of rock-crystal; which is most common along the sea-shore, and on the shores of the lakes. Sand is very extensively used in the preparation of lime-mortar, in manufacturing glass, and making water-filters. Sand, containing a small portion of alumina, is used in foundries for making moulds. We find valuable metallic substances, such as gold, platinum, copper, iron, &c. mixed with, and imbedded in sand and gravel.

SANDSTONE.

Sand, cemented together, forms sandstone. Sandstone is found in extensive mountain ranges, in heavy masses and layers. As a building material, it is judiciously and extensively used. The sandstone of the eastern slope of the Alleghenies, is a superior material for architectural purposes, being dry, hard, and capable of resisting, to a great degree, the influence of the atmosphere. The sandstone of the western, or bituminous coal fields, is not so good; it is more brittle, and liable to shrink, and, under the influence of rain, and the heat and cold of the atmosphere, it soon crumbles into sand. Old sandstone, or that nearest the primitive rocks, is the most durable, while that of recent origin, found in the tertiary rocks, is a badly-cemented sand.

SLATE.

This is an extensively-distributed mineral, forming, like sandstone and the other rocks, whole ranges of mountains. We shall here speak of but one species of this rock, having treated of the others previously.

Roofing-slate, or clay-slate, is one of the most valuable of this series; a fine quality of it is found in Eastern Pennsylvania. This slate is not discovered in secondary rocks, and is vainly sought in the bituminous coal region. The colour of roofing-slate is generally grey, or bluish-grey, inclined to black, brown, or grey, and exhibiting on its surface a red film of oxide of iron. A good slate of this kind splits straightly, into thin laminæ, and does not absorb water: this is tested by weighing it both before and after its submersion in water. It should be sound, compact, and uniform, showing no fractures or hard veins.

SOAPSTONE,

French chalk—is a greyish-white, but often greenish mineral, and is extensively distributed on the eastern side of the Allegheny mountains. It is abundantly found in Maryland, of an exceedingly good quality, where it is manufactured into various articles used in the arts. It is an excellent fire-proof material, and is used, with admirable success, for dusting the faces of moulds in iron foundries, imparting a smooth and sharp face to iron castings.

SULPHUR.

There are no minerals, containing pure sulphur, found in the United States; at least, not in sufficient quantity to be of practical use. The chief sources from which sulphur can be obtained, are the sulphurets of the metals, which we possess in great abundance. Sulphur may be extracted from iron pyrites, by simple distillation in iron or stone retorts, when they yield one-half the sulphur they contain; the remainder, sulphuret of iron, is easily converted into copperas.

TRIPOLI,

Rotten-stone—is a soft, friable mineral, of an earthy fracture, in colour yellow-grey or dirty-white, and does not adhere to the tongue like clay. It is composed entirely of silica, and is an aggregation of the skeletons of small animals.

TURF,

Peat—is a material found in low grounds and swamps, and, if sufficiently pure, forms a fuel of considerable value. Turf is an aggregation of decayed or carbonized plants, which grow and sink on the very spot where it is found. The United States do not afford much turf, and, if there were inexhaustible stores of it, it would not be much used: mineral coal, being found everywhere, can be mined and sold at such low prices, as would render peat bogs unprofitable. Turf is peculiarly adapted for heating and hardening steel, and for this purpose the turf is pressed and charred, the fire of raw turf being found injurious. Turf is found on Long Island, New York, and in other places along the sea-coast; also in Michigan, Missouri, and other States of the Union.

PART II.

ASSAYING OF MINERALS.

UNDER the term, assay—to assay—we understand particularly, the operation by which alloys of the precious metals are separated from them, and the gold and silver purified. We do not mean to allude to that operation, but to the art of resolving a compound mineral into its constituent parts. For our purpose, analysis, or chemical analysis, would be the proper term; but the term, "assay" is in such general use, that there is no impropriety in applying it.

The assay of a mineral substance may be accomplished by smelting, or the dry analysis; or by means of solvents, denominated the humid way, or the wet analysis. There is another distinction of analysis, which shows only the kind of matter of which a substance is composed, and is known as the qualitative analysis; and one which determines not only the kind, but also the quantity of each kind of matter, is called quantitative analysis. It is not within the limits of this work, nor is it the object of the author of this book, to go into the extensive details of a chemical analysis of minerals; our object is, to give such information regarding the assay of common minerals, as will enable the examiner to decide what kind of mineral is the subject of his investigation. We intend to enable the farmer, mechanic, and others, who cannot pay as much attention to this subject as a chemical analysis requires, to determine the character and value of the minerals which they may examine. It is frequently impossible to decide on the kind and quality of the mineral under observation, by the mere application of our senses to the raw material; we are often compelled to subject the mineral to some artificial treatment, in order to accomplish our purpose. As we progress in the examination of a mineral, it should be subjected to the various trials which we are enabled to make by the use of our senses, to the alteration of its form by heat, or moisture, or such other means as will prove the most effective; and such apparatus and ingredients should be applied, for this purpose, as may be found without much difficulty in every house.

EXAMINATION OF WATER.

If the substance under examination is a fluid, say water, or water mixed with earthy matter, smell and taste are, in this case, powerful agents in detecting substances dissolved therein. Pure water is free from taste and smell, and has a cooling effect on the tongue and stomach. Rain-water has a peculiarly soft taste, resembling weak ley. In case the taste is not sufficient to decide whether spring water be soft, or, in other words, pure, it may be effected by pouring some clear soap-water into it; if it become curdy or turbid, we may conclude that the water is hard, not suitable for washing, and contains some foreign substance in solution.

Spring water, well water, and all the water extracted from the interior of the earth, is more or less adulterated with saline matter. A small quantity of foreign matter imparts to water, when used as a beverage, a cooling and refreshing effect, like that of the common water of springs and wells.

Mineral water.—If the amount of foreign matter is increased to a certain extent, or is of a certain kind, we call the water mineral water. The taste of mineral water is not generally agreeable, there being but a few kinds which taste well, and produce a pleasant sensation when drank. All well and spring water is a species of mineral water. If water contains more foreign matter than is compatible with our sense of taste, it is called acidulated water, or brine.

Water which is of a crystalline purity, and has an agreeeable and refreshing taste, generally contains pure carbonic acid gas. If we cannot decide on the presence of this gas by the taste, we may detect it by allowing a tumbler, full of water, to stand in a warm room for a few hours, when the inside of the glass will be covered with small bubbles of this gas; the more gas the water contains, the greater will be the quantity of bubbles found adhering to the glass.

To decide positively whether there is any carbonic acid gas in water, we must pour into it a little clear vinegar, and stir it well; after this, pour in a little finely-powdered sugar, when the gas will rise rapidly, in small bubbles, and, if there is much gas, it will occasion an ebullition. By inhaling the liberated gas through the nostrils, it produces an agreeable and refreshing sensation.

Water possessing the smell and taste of rotten eggs, is hepatic or sulphurous; this peculiar smell is produced by sulphuretted hydrogen gas, generated from soluble metallic sulphurets. If the quantity of sulphuretted hydrogen is not too large, the taste of this water is not repulsive, but the smell is decidedly so. Waters of this description possess remarkable medical properties, and are famous as resorts for invalids.

If water has a cool taste, but has an additional earthy after-taste, it contains lime; either sulphate of lime, gypsum, or carbonate of lime—dissolved limestone. Magnesia causes water to have a bitter taste, resembling that of Epsom salt, which is not unpleasant, if the quantity of matter dissolved is not too large. Solutions of iron impart to water a taste like black ink, being very disagreeable if there is too much of it in the water. Potash and soda do not give very distinct tastes to water, having more the taste of their acids than of their bases. These salts appear chiefly as chlorides in mineral water, producing the peculiar taste of common salt. If the sulphates of potash or soda are present in water, they impart a bitter taste, very much resembling magnesia, but stronger and more repulsive. Some waters contain clay in solution, but always accompanied by a strong acid; they have a decidedly acidulous taste. It is necessary to use other means than the tongue, to enable us to distinguish the acid in these waters.

Saline waters are more impregnated with foreign matter than mineral water, are repulsive to the taste, and have a strong medicinal effect. It is easily understood that saline waters may either contain many ingredients, or but one kind of matter; the first is most commonly the case. Saline waters are chiefly those which contain a considerable amount of chloride of sodium, as, for instance, sea-water and brine-springs. They are also found to contain sulphates, as predominating elements, which are easily discovered by their bitter taste.

Turbid water.—If water is turbid, and we wish to know the amount and description of solid matter which may be held in solution, the best method of ascertaining is, to fill a vessel with the water, let it remain undisturbed until all the suspended matter has settled, then gently pour off the clear water from the precipitate, which may be dried slowly, and tested in the same manner as any other mineral. The sides of the vessel should be perpendicular, or, what is much better, the vessel should be narrower at the top than at the bottom; in that way, the sediment settles sooner, and the water will be clearer when poured off.

Such sediments, next to those of alumina, vegetable matter, iron, &c. consist generally of siliceous matter. The sediment, possessing generally the appearance of mud, should be put in a white porcelain dish, and dried by a gentle heat. If it shows a strong tendency to contract, while drying, separating into small, isolated flakes, we may conclude the sediment contains alumina; and if the flakes shrink a great deal, it contains proportionately more alumina. Silica does not shrink at all, but forms a continuous covering over the plate; lime and other matter act in a similar manner. A dark-coloured sediment indicates vegetable matter, which may be proved by sprinkling some of it on a red-hot iron; if it loses its dark appearance, we may assume that the colouring is caused by the presence of vegetable matter. If the colour changes to red, the sediment then contains iron. If we wish to ascertain the amount of solid matter in a certain quantity of water, the water should be measured or weighed, and the gently-dried sediment being scraped off the plate, will give, when weighed, the portion of solid matter contained. One gallon of water usually weighs eight pounds.

Solution of salt.—If soluble matter, which cannot be obtained by precipitation, is contained in water, or there is both soluble and solid matter in it, then the water should be evaporated over a gentle fire, or, what is still better, in a warm room, by spontaneous evaporation. The soluble matter will then crystallize, and be found in the bottom of the dry vessel in the form of small crystals, which are often very fine, and feel gritty, like sand. These crystals may be tested in the same way as other matter, and may also be weighed, to find the quantity contained in a certain amount of water. Copper, porcelain, or glass vessels are necessary for this purpose.

EXAMINATION OF SOLID MATTER.

The examination of solid matter is not so easily accomplished as that of water, owing to the greater variety of substances coming under this head. Minerals proper, are seldom or never distributed in large bodies, seldom forming ranges of mountains, or even single mountains. The latter happens with iron, but with hardly any other mineral. If we find a substance in large masses, we may approximate to its value, until properly examined. With the exception of iron, minerals are either scantily dispersed in sand or soil, or exist in small veins, traversing and ramifying rocks. All masses of minerals are, either granite, sandstone, slate, or limestone; and, however valuable these may be, it is not our province to examine them.

Minerals imbedded in sand or soil.—Minerals, or ores of metals, are generally of greater specific gravity than soil or sand, and this forms the principle upon which to base the first examination. If sand or soil is to be examined, we must select a spot where we think minerals may be found. This may be on the bank or in the bed of a river or creek, in the pool of a rivulet, or in a drain or ditch in the field, or, in fact, in any place where a current of water would be likely to deposit a heavy substance. If about to examine sand, clay, or soil, containing such minerals, we should put it in an iron or tin pan. A suitable one for this purpose must be but ten or eleven inches in diameter, and from two to three inches deep; the bottom of the pan must be straight, and not be either concave or convex, the brim terminating in a sharp angle with the bottom, so as to form a well-defined corner. A common sheet-iron or tin pudding-pan is best adapted for this purpose. Such a pan should be filled with the sand under examination, and then carried to a basin filled with water, standing either in a large tub, or in a pool, in a creek where there is no current. Washing under a pump, hydrant, or in a current of water, is not recommended; for, some of the matter which we might wish to save, might be carried off. The pan, filled with sand, must then be so far submerged, as to saturate the sand fully, and fill the pan with water. The sand must now be stirred with one hand while the other holds the pan, and when the mass is well worked through, the muddy water in the pan must be poured off, the pan again filled with fresh water, re-stirred, and poured off. This process must be repeated, until all the fine, light particles of clay and sand are washed off. If there are any lumps of clay among the sand or soil, they must be carefully rubbed and dissolved, so as to avoid washing away any particles which may be enclosed in the clay. When all the fine substances are washed away, which will considerably diminish the bulk of sand, the pan must be held horizontally, and only so much water allowed to enter it, as will cover the sand. By holding the pan with one hand, and shaking it with the other, all the heavy particles will sink below the sand. Gradually lowering the side of the pan which is shaking, that is, inclining it towards the hand which is in motion, will allow the light particles, if even large, to flow off with the water, and the heavy metal will settle in the corner of the pan, from which it is with difficulty moved. If there is much quartz sand in the pan, covering the particles in a thick layer, the greater part of it may be drawn out by one of the fingers, care being taken, at the same time, not to throw out any mineral. It is not necessary to use both hands in shaking the pan; a little practice will enable the operator to shake with one hand and wash with the other.

In this way the quantity of sand is reduced to a minimum; it may be washed off almost to the last grain, if the shaking operation is well performed. In the corner of the pan there remains more or less metal, or metallic ore, in case there was any in the sand. By putting a small quantity of water in the pan, just enough to cover the sediment, and turning the pan round in such a way as to produce a gentle current in the corner of it, the minutest particles of metal, even one particle, and that invisible to the naked eye, may be secured. If there are any particles heavier than those which are visible, they will be brought to light, because the lighter particles are carried forward by the current of water, and the heavy ones remain behind. In the corner of the pan will now be seen a string, or some one or more heavy particles, which may be examined by a lens if they are too small to be detected, or are forms which cannot be distinguished by the unaided eye.

In this manner sand or gravel is washed, to obtain gold, platinum, diamonds, tin, and lead. Gold is easily detected by its bright lustre, platinum by its lead colour, and by its weight, tin by its dark grey, often black colour, and lead by the lustre and crystal form of the galena. Other metals can hardly be distinguished from each other in that way, because their sulphurets are easily oxidized, and escape in this way without being recognized. Diamonds are discovered, by holding the pan in the direct rays of the sun. Materials which cannot be distinguished by the eye, are subjected to those trials which are described in the course of this part.

Minerals from veins or beds must be cautiously selected, so as to have the same kind of material to examine. In a vein of mineral belonging to the primitive or transition rocks, there are always a variety of substances; but in the coal series much variety cannot be expected. If a small amount, say a handful of minerals of the same kind, are gathered, they are subjected to an examination by the eye, and if necessary, aided by a lens; the colour, lustre, and crystals are observed; and if we cannot decide by the sight what the mineral may be, we try it by weight, and approximate to its specific gravity. If a handful of the mineral has twice the weight of quartz, its specific gravity may be estimated at 4½ or 5, because that of quartz is 2½. Experience in judging of specific gravity by feeling, is of great service, and vastly facilitates the examination of minerals. The specific gravity of water is 1; common brick, 2; quartz, 2½; sandstone, 2 to 3; most of the compact iron ores, 3 to 4; heavy spar, 4½; tin ore, 6; sulphate of lead, or galena, 7; gold, 19, and platina, 21. From this we may extract some leading features, which can be used as a standard of comparison for other materials. When the eye, tongue, smell, and gravity have done their part, the hardness of the mineral must then be tested. It must be tried successively by the finger-nail, an iron nail, a knife, a file, or the hardest steel that can be obtained; and the mineral having been scratched by each one of the above enumerated articles, may be attached to a certain class. This last examination should be very cautiously made, and much dependence cannot be placed upon it for the accomplishment of our purpose; the same mineral being often found varying in hardness.

If the mineral under examination is white, or its predominant tint is white, and it is also soft, it may be either clay, chalk, limestone, or some metallic oxide. It should be gently dried, and tested by the tongue; if it adheres to the tongue, it certainly contains a portion of clay. This test may be corroborated by breathing upon it; if it emits a peculiarly repugnant odour, it may be considered as clay, or a mixture of clay and some other substance. If a mineral is found in the lowlands, or in a saddle on a hill-side, we may assume that it is common clay, that is, a mixture of silex, or sand, and alumina. Clay is also found in veins, in the rocks of all ages. If the mineral is chalk, or a metallic oxide, and we know the locality whence it is derived, we may easily decide its character. Chalk does not appear in veins, and metallic oxides and other minerals are not, like chalk, found in extensive mountain ranges. If we are not acquainted with the locality from which the mineral is brought, and it possesses the appearance of chalk, it should be first exposed to a gentle heat, and, afterwards, heated to a cherry-red; if it still continues to retain its white colour, and slacks in water, like quicklime, we may say that the mineral is chalk. If it is neither clay nor chalk, it may be a metallic oxide. We should now proceed to the operation of roasting, and examine it as will be explained hereafter; for, after roasting, minerals are reduced to more simple elements, and may be subjected to a general examination.

If the mineral is hard, but still white, we may test it by the tongue; if it adheres to the tongue, it may be clay-slate, clay, fire-clay, or argillaceous iron ore.

It should be pounded and roasted, after which the quality may be ascertained. If it does not adhere to the tongue, it may then be a metallic oxide or carbonate, for most of the carbonates are white. If the mineral has a crystal form, it requires an expert mineralogist to decide on its classification, and even then the result is very uncertain; the only positive method of deciding the question in this case is, by roasting the specimen. If the mineral is so far soluble, as to impart a taste to the tongue, it may then be common salt, which is easily distinguishable; saltpetre, possessing a disagreeably sweet taste; alum, which is sour, and has a strong taste of iron; or white copperas, which is sour, and very much like ink in taste.

If the mineral is yellow, it may be, and in most cases is, hydrated oxide of iron, particularly if it is friable. A few lead ores are also yellow, but they are regarded as mere curiosities. Zinc has frequently a yellow appearance, but the colour is never very distinct, being always dirty; it is of a compact, aggregate form. If the mineral has a yellow colour, is hard, crystalline, and of a metallic lustre, it is, probably, a metallic sulphuret; this can be determined by the process of roasting.

If the mineral is red.—This embraces a very extensive range of minerals, in which iron predominates in quantity. There are masses of red iron ores of every variety of form and colour, from a faint, rose-coloured clay-ore, to the dark crimson, almost black, crystalline oxide. Most of these ores adhere slightly to the tongue. Iron ores are generally found in large masses; and, if the mineral is found in that way, it is a strong indication of iron. If the sight is unable to decide on the ore to which the specimen belongs, we have no other resource remaining but to roast it. Red zinc ore is distinguished from the iron ores by having a shade of yellow in its composition, while iron inclines to brown. Cinnabar is also red, and, like zinc, inclines to yellow or orange; but it is distinguished from zinc by a more lively colour, and from iron by its shade of colour. There is also a red copper ore, which is of a bright fiery red, and has a greater specific gravity than the above-mentioned ores; it is in many cases difficult to decide upon its character, as it is very seldom so free from foreign matter, as to afford an opportunity of comparing its specific gravity. Lead ores, if red, are characterized by a dull colour and great specific weight, which is, in the plurality of cases, twice that of iron ore, and in every instance, at least one and one-half times its weight. There are also red silver ores, red ores of antimony, and other red minerals; but these are substances which should, properly, be left to the examination of a professional mineralogist.

If the mineral is brown.—This colour predominates in the mineral deposits. We have here the compact hydrated oxide of iron, which is brown, and has a glassy, earthy, and metallic lustre in the fresh fracture. Where this ore is exposed to the action of the atmosphere, it is usually covered by a film of yellow hydrate. This brown iron ore yields a yellow powder, by which it may be distinguished from other minerals. The brown ore of antimony is the most common ore of that mineral; it is the sulphuret of antimony, easily recognized by its well-developed crystals, or by its crystalline fracture; the crystals are developed in long prisms or pyramids. The colour of this ore always inclines to blue. The common tin ore—oxide of tin—is also brown verging on black; it has but little lustre, is very hard, and extremely heavy. The sulphuret of zinc is brown, ranging from light brown to a dark, almost blackish brown, and it is so soft as to be easily scratched by an iron. This ore is distinguishable by a slight transparency at the corners, and on the sharp edges. Brown cinnabar is that species of sulphuret of quicksilver which is contaminated with any vegetable matter, carbon, or bitumen; it is quite common among the ores of mercury. All the ores of manganese are brown, more or less inclined to black; they are found crystallized, and amorphic, in masses of earthy texture. There are quite a variety of lead ores of a brown colour, but they are usually accompanied by galena, and may be easily known; these ores are seldom found in such masses as to afford any profit in mining them alone. Chrome ore is also brown, but not easily distinguishable from the ores of other metals.

If the mineral is black.—Here we again have iron

as the most prominent in the whole range of minerals: the immense masses of magnetic ores are all black. They are distinguishable from all other substances by their affinity to the magnet. If a magnet cannot be readily obtained, a well-cleaned knife, rubbed strongly on the powder of the ore, will attract it. This mineral has always a tendency to a bluish-black colour, is exceedingly hard, and, in the majority of cases, crystallized. A velvet-black hydrate of iron is also found, but always accompanying the brown hematite; it appears in beautiful concretions. A crystallized, black oxide of iron, having a feel like plumbago, is very common. A black ore of copper is also found, but it is a curiosity, generally forming but a film of black velvet oxide over another ore of copper. Most of the common manganese ores are black, but usually inclining to a brown, and, in some instances, to a blue. Metallic silver appears also in black masses, or concretions, of the greatest variety; it is malleable, and, when cut with a knife, exhibits metallic silver.

Lens.—In conducting examinations by the eye, it is necessary to have the use of a good lens—magnifying-glass—as an assistant. A common single lens, one inch in diameter, answers very well; but it is much better to use a double glass, that is, a lens with two glasses. The common magnifiers, of three-fourths of an inch glasses, are rather small for the examination of minerals; they require glasses with a large focus. For these reasons, magnifiers ought not to be less than one inch in diameter; and they will be much better if as large as one and one-half inches. In examinations by the aid of the lens, the mineral should be broken, to afford a fresh fracture; if, by close examination no difference in the texture, or sprinkling of foreign matter can be detected in the mass, the mineral may be considered of uniform composition. If grains or crystals of other minerals are found imbedded in the main mass, they should be examined as to their colour, lustre, hardness, and crystal form, and from this examination a deduction may be made as to the nature of those mineral particles. If we cannot in this way decide on the quality of the mineral, the whole mass should be carefully pounded and washed; we will thus obtain a larger quantity of particles, which may be more readily recognized. If it should not be in our power to distinguish the mineral after pounding and washing, there is no other plan left but to roast it. In examining minerals by the eye, or by the help of a lens, it is advisable to moisten the specimens, to draw out more vividly their colours and lustre. The specimens, however, ought not to be wet so much as to increase their lustre, and diminish the sharp expression of the crystal.

Examination of gold ores.—When gold ores are under examination by the aid of a lens, the greatest caution is requisite not to decide, until the mind is perfectly satisfied in every particular. Gold ores, containing gold in very minute particles, are the most difficult to examine; a speck of iron—sulphuret or oxide of iron—will frequently lead the best observer astray. If, apparently, a speck of gold is detected in a piece of rock, the specimen should be turned in such a way, that a direct light may fall upon the grain or particle from all directions, and it should then be minutely inspected in every possible way. If we find, on turning the specimen around, no portion of the surface, where colour, curvature, or lustre is interrupted, we may, provided the specimen be moist, believe the speck to be gold. If the particle exhibits a perfect plane, or a sharp angle, we may question its identity with gold; still, it may be gold, which can be decided by touching it with the fine point of a knife. A metallic sulphuret will not take an impression from steel, but gold will. If the sight cannot decide whether the specimen does or does not contain gold, the surest way is, to pound the rock into a fine powder, carefully wash from it all the rocky matter, and then examine the pan, if necessary, by the assistance of a lens, to ascertain if there be any particle of gold in the corner of the it. Particles of gold exhibit a uniform colour, and are either flat spherical spangles, or round grains; while particles of sulphurets, or other minerals, have crystalline forms, refract the light more strongly from one side than from the other, and also show planes and angles on the surfaces.

The forms and other appearances of minerals are

described in the first part of this work, and we beg leave to refer the reader to that, for any information he may require in addition to the foregoing.

Roasting the mineral.—If it should be beyond our power to determine the nature of a mineral by its exterior qualities, we should then proceed to roasting or burning it. The most convenient plan is, to take one or more pieces of the mineral, and expose them, for at least twenty-four hours, to the heat of a common grate-fire. A heat for roasting should never be higher than a cherry-red, and, if it can be maintained below that degree, it is advisable to do so. A heat sufficient to melt and burn sulphur, is amply sufficient for the roasting of any ore; a greater heat is useless, and, in most cases, injurious to the process. Extended time, and a low heat, are the most advantageous for this process. Minerals, which are not sulphurets, and even these, if in large crystals, are liable to fly into pieces, when brought suddenly in contact with the fire; it is advisable, in order to prevent this, to lay the specimen near the fire to dry, for a period of twenty-four hours or longer; for, the enclosed moisture, being suddenly heated, causes the mineral to fly. After a mineral has been exposed for one or two days to the influence of a gentle heat, it changes its appearance, becoming more friable and porous, and, in most instances, changing its colour. If the heat should be too powerful for certain minerals, they will partially melt, and it is then impossible to roast them, as they will not assume that open, porous form, or become more highly oxidized. In this case there is no remedy, and there is no way left to accomplish the desired end, but to pound and then roast the mineral again.

Pounding the mineral.—The pulverizing of rocks or minerals is a simple operation, if there is nothing else required than the conversion of the whole into a fine powder. This, however, is not always the case, for the mineral we wish to examine is frequently found in but small grains or crystals, interspersed in the rock, and we want not the rock, but merely these grains. In this the pounding is more difficult than in the first case; it is best performed in an iron mortar; but, if such a one cannot be readily obtained, the pounding can be performed upon a large, hard, and smooth gravel-stone, either with a hammer, or another hard stone as a pounder. Around and below the stone should be laid a large sheet of paper, or a cloth, to receive the scattering pieces. If the mineral is to be but coarsely pounded, so that the rocky matter may be washed away in the pan, the mineral should only receive gentle, perpendicular blows, and all rubbing should be avoided. If we wish the mineral entirely converted into a fine powder, this may be partially and finally accomplished by rubbing. When the mineral has been sufficiently pulverized, we may either wash the light, rocky matter off, by means of the wash-pan, and examine the residue with the lens, or, if we do not succeed in determining on the nature of the material, we may continue pounding and prepare for roasting. If there should be but a small quantity of ore in a large piece of rock, it is advisable to concentrate the ore by washing off the rocky matter; in fact, it is never hurtful to the pounded ore to wash it gently, and rid it, as much as possible, of foreign matter, previous to the operation of roasting. Foreign matter, particularly silex, which is the most usual adulteration, may be easily detected by the aid of the lens. Clay, limestone, and similar substances, when mixed with the ore, may be pounded very fine, and then readily washed off. If the material to be pulverized, is of such a hard nature as to make it difficult to pound, then the specimens may be heated to redness and suddenly plunged into cold water; if one trial does not produce the proper effect, the heating and chilling may be repeated, until the mineral is sufficiently pulverized. If a mineral is too soft or unctuous, and will not form a sufficiently fine powder for our purpose, it may be mixed with fine, white sand; its bulk and white tint are the only disturbances to the examination, the whiteness of the sand diminishing the colour of the mineral.

After pounding and washing, the mineral may be again examined by the eye and lens, and, if its nature cannot be ascertained, it will be necessary to roast it. This can be best performed in an earthen-ware dish, and, if no such dish can be procured, it may be done in a cast-iron pot, or any kind of flat, iron vessel; cast iron is, for this purpose, preferable to wrought iron. The powdered ore should be put into a flat vessel, as above-described, and placed upon a fire, where it may be exposed to a cherry-red heat. The heat may be gradually raised, while continually stirring the powder; care being taken never to allow the heat to be greater than a cherry-red. With some ores this operation requires but a few hours, while with others it cannot be accomplished in a whole day.

The principle involved in roasting ore is, to oxidize the specimen highly, or evaporate the water, sulphur, arsenic, carbonic acid, or carbon, and combine the metallic elements with oxygen, which is always accomplished by the highest degree of oxidation. If we should expose black magnetic oxide of iron to the heat of a coal fire, it will merely be oxidized to a much greater degree, and form red per-oxide of iron. If iron pyrites should be roasted, the sulphur would be evaporated, and the iron basis absorb so much oxygen, as to form per-oxide of iron. When arseniuret of iron is roasted, the arsenic is generally evaporated, and the iron remains oxidized to such a degree as to equal per-oxide of iron. Carbonate of iron, both the sparry and the compact, requires a great heat to drive off all the carbonic acid, but the iron always remains in the form of a red per-oxide of that metal. There are but few exceptions to this rule; all the minerals with a metallic basis oxidize, while roasting, to their utmost capacity. This law forms one of the most significant features of the roasting process. For these reasons, it is necessary that the minerals should be as pure as possible, and that the roasting should be performed with care and attention. Should a variety of metallic ores be mixed in the same roasting, the colour resulting therefrom cannot be considered a criterion in determining the kind of ore under treatment. If the heat was too great during the roasting, and the ore has partially or entirely melted, the appearance of the roasted ore will be quite different from that which has been properly roasted. By roasting, we mean, heating a mineral to redness, under a liberal access of atmospheric air. If, therefore, we wish to be successful in this operation, this condition must be complied with, and all coal and other impurities, carefully excluded from the mineral.

If the mineral should be a sulphuret, a low degree of heat will be sufficient to liberate the sulphur. If a piece of hot sulphurous mineral be placed close to the nose, so that we can inhale the vapours emitted by it, a more or less strong, suffocating odour of sulphurous vapour will be received from it. Carbon contained in a mineral will burn out without the emission of any odour. Arsenic, antimony, and zinc, do not generate any smell, but throw off white fumes, which may be condensed on a piece of cold iron held over the heated mineral. These vapours are all very poisonous.

One of the principal conditions necessary to the exposure of all particles or atoms in a roasted mineral is, the access and influence of the atmosphere; it is, therefore, essential, that every precaution should be taken to prevent the mineral from melting, it being impossible for the atmosphere to penetrate a melted body. This condition is not easily complied with; for, sulphurets usually melt at a low heat, as do also arseniurets and combinations of antimony. The worst minerals in this respect are the impure iron ores, such as the compact carbonates, magnetic oxides, and silicates: if these ores once melt into a black slag, they are past redemption, and it would cost hard labour to roast such melted iron ore. Sulphurets, and all other ores, if once melted, may still be oxidized and perfectly roasted; they should be, for this purpose, finely pounded, and again roasted; if there remains any doubt of their perfection, they must be once more pounded and roasted. This process may be repeated as often as may be considered necessary to obtain a good result; when sulphurets of lead and zinc are mixed together, this is a very tedious process, as they occupy a long time in roasting, and must be repeatedly pounded.

Ores which fuse readily roast with difficulty. To facilitate the operation, the mineral may, when finely pounded, be mixed with some other substance, which will prevent the ore from melting; or, if it does melt, the substance enclosed in the melted grains, or clinkers, will open them again, on being stirred over a lower heat. Common salt, saltpetre, and other ingredients are used for this purpose, which partly assist in the process of oxidation, but are chiefly useful in interposing their particles between the particles of the mineral, and preventing their contact. In our case, we desire only to bring the oxidation of the ore to a higher, and, if possible, to the highest attainable degree. It makes no difference to us if there be foreign matter among the ore, provided that matter is white. Saltpetre, salt, and other compounds of the fixed alkalies, frequently form compounds with the ore, which have a tendency to spoil the colour, and so defeat our object. In this case, it will be advisable to take very pure, white sand, and pulverize it along with the ore ; an equal weight, or more, may be used to advantage. Roasting white sand or quartz along with an ore, generally develops the colour better, and brings it to a more decided expression, than can be done by the use of any other substance.

In some instances it may be necessary to add charcoal powder to the finely-pounded ore; this will be the case where a combination of arsenic is to be dissolved. Arsenic adheres closely to other metals, but if we add a little powdered charcoal to the roasting ore, more arsenic will be driven off than would be the case if the coal were not used.

If correct results are expected from the process of roasting, it is necessary that the ore should be pure, that is, all of the same kind. An addition of sand or quartz does no harm, but is, on the contrary, beneficial. The heat should be as low as possible, and never beyond a cherry-red. If the mineral is in powder, it must be continually stirred; and if there is any indication of melting, it must be pounded and again roasted.

Distinctions between roasted minerals.—If the roasting operation is well performed, we will find the metallic ores of such distinct colours, that we cannot fail in their classification. The ores of IRON, if pure, are all turned crimson-red, frequently, very dark, if the mineral is remarkably close; the more porous the mineral has been, the lighter will be the red, the more of a cinnabar colour will it assume. Very porous, and purely yellow hydrated oxides of iron frequently assume a most lively, beautiful fiery red. This red oxide of iron is not attracted by the magnet; but the roasted specimens are seldom so well oxidized as to be free from particles of magnetic ore, which are, of course, attracted. The red colour of the iron oxide is not peculiar to iron, other metals form oxides of a similar colour; but the iron may be distinguished by having a dull colour, while other oxides are more lively. Iron ores may be roasted most conveniently in pieces, on a grate or in a stove, or on any fire which will not melt it. If in pieces, it will require at least twenty-four hours, but better if forty-eight hours, before it may be considered properly roasted. The sparry carbonates require a strong heat and a great length of time to oxidize them thoroughly.

Copper ores, when roasted, are black. Copper ore forms a black oxide. If pure, it has a velvety colour, but if it is impure, the colour will be more or less modified. An addition of iron gives it a brown appearance; most of the other metals impart to it a shade of blue.

Zinc generates a white oxide, which, if mixed with other minerals, as is usually the case, is not easily recognised. If, while the ore is roasting, we add a little powdered charcoal, and apply a stronger heat than usual, flowers of zinc will be formed on a piece of cold iron, held over the roasting-pot. These flowers, which are composed of oxide of zinc, can be readily distinguished from other white fumes or metallic oxides; the white oxide of zinc being the only one which forms a kind of organic crystallized matter, resembling snow-flakes. If there is but a small quantity of zinc in a mineral, it will be difficult of detection in this way; and, even though one-half the weight of the mineral should be zinc, it will be unsafe to decide by this test.

Lead produces a yellow oxide when roasted. This colour, however, cannot be depended upon; for, if the lead should contain other metals, the colour will incline to a green, brown, grey, or a more or less dark, uncertain stone-colour. It appears also, in the well-known colour of minium, produced from galena, when roasted by a low heat. Lead, particularly sulphuret of lead, is of a very volatile nature; therefore, we should carefully avoid the use of a great heat in roasting it. Lead ores appear in such a great variety of forms, that it is almost impossible to decide, by the colour of the oxide, whether the mineral under examination is, or contains lead. The safest characteristics to depend on, are, the heavy weight and great specific gravity of the oxide, which is greater than that of any other oxide. Oxide of lead is very fine, and contains no grit; other oxides are more or less gritty. It melts at a lower heat than any other oxide.

Antimony is commonly found in the form of a sulphuret, which is very volatile, easily fused, and requires great caution in roasting it. The sulphur cannot be entirely separated by roasting; and this causes the roasted matter to have a dark grey colour, a mixture of sulphuret and protoxide. The more perfectly the desulphuration of the ore is performed, the more will the colour approach to a faint yellow. Impure lead ore has this characteristic in common with antimony; but, before roasting, it is not difficult to decide between antimony and lead.

Tin, when properly roasted, forms a dirty yellow powder, containing sesquioxide of tin. If but imperfectly oxidized, it forms a grey oxide of tin — putty of tin. Oxide of tin is remarkably heavy, almost as heavy as oxide of lead; it is distinguished from the latter by its containing a larger amount of gritty matter.

Quicksilver. — Cinnabar cannot be roasted; it is so volatile as to evaporate with the sulphur which may be liberated from it. The best plan for testing quicksilver ore is, to mix the pounded ore intimately with some dry quicklime, place them together in an iron pan, and apply a gentle heat; after the heat has been applied for a short time, a globule of mercury will be found in the bottom of the pan.

Gold.—Iron pyrites form the matrix of gold; it is enclosed in the pyrites, and these must be opened to liberate the gold. To test the presence of gold in a specimen of pyrites, it should be exposed in a solid piece, for the space of two or three days, at least, to a gentle heat. Great care should be taken not to have too great a heat, and to allow a liberal access of atmospheric air. If the proto-sulphuret, which forms soon after the ore is exposed to heat, is once melted, it will take a long time to oxidize the mass. In that case, the best plan will be, to pound the melted sulphuret, and roast it in an iron pot; this process is, however, a very tedious one. The roasting must be performed with particular care; for, the particles of gold are very apt to enclose themselves in particles of sulphuret, and thus escape detection. *All* the sulphuret should be converted into oxide, or the gold cannot be separated from it. The oxidized iron must be washed like any other gold ore, the manner of doing which has been previously described.

The other metallic ores are not found in a sufficiently pure state, to enable them to afford a characteristic oxide, and allow us to point out distinctive marks. The only way to arrive at a satisfactory result is, to have such ores analyzed by a professional chemist or metallurgist.

Assay of ore by smelting.—It is not our object in writing this treatise, to describe the operation of assaying so perfectly, as to afford a qualitative and quantitative result, equal to a chemical analysis. We intend to go as far into the subject as will enable the operator to form a tolerably correct estimate of the value of the ore under treatment. The operation to be performed is, to reduce the ore to its metallic state, and then decide its value. To perform a dry analysis or assay of this kind, crucibles are required for the smelting of the ore, and cupels, for ascertaining the presence and the amount of the precious metals.

Crucibles can be most advantageously purchased in a drug store; but if they cannot be obtained in that way, we must resort to manufacturing some. Crucibles suitable for our purpose should be from three to four inches high, and three inches wide at the top. A good-sized, well-tapered tumbler, is of the size proper for a crucible. The material of which a crucible or melting-pot is generally made, is fire-clay, which contains a large portion of sand. In many instances, an iron pot may serve the purpose. The first step necessary for the formation of a melting-pot is, to find a white, tenacious, plastic clay, which can stand the fire without melting. In the next place, some old fire-brick must be obtained; and, if these cannot be found, pieces of white porcelain, Chinese ware, or grey-stone ware, must be used in their stead. These must be pounded into a coarse sand. If these articles are not obtainable, white quartz, pebbles, or pieces of stone, are substituted.

They are heated to redness, and suddenly thrown into cold water, after which they are pounded, as stated above, and are then ready for use. This coarse sand, made from one or all of the ingredients mentioned, is mixed with as much fire-clay as will make it adhere together. Too small a portion of clay has a tendency to weaken the crucible, and too large a quantity makes it liable to form cracks and pores. The mixture of clay and sand must be well worked by hand; it cannot be too much kneaded, if we wish it properly mixed. This mixture must be coated over a glass tumbler, and laid on to the thickness of one-fourth or three-eighths of an inch, if the crucible is intended to be large; but if a small one is desired, a smaller quantity will suffice. If the glass tumbler is of a large size, and small crucibles are required, the clay may be put on the inside instead of the outside of the tumbler. Instead of a glass tumbler, a pattern, made of wood, earthenware, or any other material, will answer the purpose equally well. The bottom of the crucible should always be a little thicker than the sides. If the clay is inclined to stick to the pattern, it may be surrounded by a sheet of paper, and the clay spread upon it, which will effectually prevent its adhesion. The paper may be left in the crucible, as it will protect the clay against the first influence of heat, and prevent it from cracking; the paper will burn out in the subsequent baking of the crucible. When the form has been given to a crucible, it should be allowed a few days to dry, in a gentle heat, as it takes some time for the water to evaporate from the clay. When the crucible has been sufficiently dried, it is then baked, that is, exposed to a strong red, or even to a white heat. The baking may be done in a stove, in a grate, or in an open fire; the crucibles should be piled in the centre, and the fuel, stone-coal or dry wood, laid around them. They should be allowed to remain in the fire until it burns out; and, if the baking has been done in an open fire, they should be well covered with ashes, to protect them from the sudden rush of cold air which will follow the exhaustion of the fire. Some small slabs may be made, and burned with the crucibles, to serve as covers for them.

Cupel. — The cupel is another apparatus for melting; it is a small, flat crucible, of from one to two inches in diameter, and one-half or three-fourths of an inch high, flat at the bottom, and having at the top a flat concavity, in which the metal is assayed. The cupel is made of finely-pulverized bone ashes, of wood ashes, or of marl. The first is the best, or the first mixed with the second. These ashes, that is, the white bone and wood ashes, must be well pulverized and sifted, and as much water mixed with them as will cause them to adhere slightly. The ashes should then be pressed into the bottom of a glass tumbler, or of any other vessel having the requisite form; a small tin cup, or a simple ring of tin being sufficient for this purpose. The cavity may be made by the hand, and forms a dish shaped like an extremely convex watch crystal. The mass having but a slight adhesion, requires gentle handling; for this reason, a paper cover may be laid around it. These cupels should be porous, but close enough to prevent the infiltration of hot metal; and, on this account, too much and too little water will prove equally injurious. A strong pressure is necessary in forming the cupel. The fresh cupels are but air-dried, which may be performed on the top of a common stove, or in any other warm place. After this, they are ready for use. The ashes made of the bones of sheep are preferable to ashes made of other bones.

Smelting.—A mineral ore must be smelted, in order to obtain the metal which may be a part of its composition. The principles involved in the smelting of a metallic ore are, to mix the ore with such other matter as will absorb that which is combined with the metal; and, also, to mix the ore with such substances as will absorb, or combine with the foreign matter in the ore, and produce a fusible slag, which is to be as fusible, or, if possible, more fusible than the metal to be produced. The degree of heat used in this process, must be high enough to liberate the matter combined with the metal, and also high enough to melt the metal, as well as the cinder formed of its impurities. The mineral, if an ore, is always a combination of metal and oxygen, or metal and sulphur; for, all other combinations are changed into the first by roasting, before they are exposed to reduction, or smelting. Oxygen readily combines with carbon in a heat sufficiently intense for smelting; therefore, all the oxides are reduced by carbon. Sulphur has a great affinity for iron; therefore, sulphurets are always melted along with metallic iron, which absorbs the sulphur, and produces the metal which was at first combined with it.

The ingredients used in smelting, as well as the ore, must be finely pulverized, so as to form a fine, impalpable powder. Ore, fluxes, and the material for reductiou, are often all mixed together; at other times they are put into the crucible separately. All the ores to be smelted must be as pure as possible, that is, free from foreign matter. The coarsely-pounded ore should, in every instance, be washed, to purge it, as much as possible, of all impurities.

Smelting Furnaces cannot be everywhere obtained. If they could, their use would be preferable to the means I intend to propose for supplying their place. It is fairly to be presumed, that those who are willing to incur the expense of building a smelting furnace, have also the knowledge necessary for that purpose, without resorting to the narrow limits of our work for such information; we will, therefore, pass that matter by, and proceed to the statement of our proposed method. For smelting lead, and almost all other ores, except iron and tin, a common stove will give out sufficient heat, particularly if the stove is provided with a clay lining. A common fire-grate constitutes an excellent smelting furnace, if the face-bars can be shut, or, at least, partially closed. This may be done by the use of common clay. The grate may then be filled in with brick at both ends, so as to reduce the interior dimensions of the grate to one foot in length. The space above the grate, which commonly furnishes the draught, may be closed by a sheet-iron door, or filled up with brick, leaving only a small opening, for supplying the fuel and inserting the crucible. This small opening may be closed by a brick. Such a grate, a pair of tongs, an iron poker, crucibles, ore, and fluxes, are the only requisites for the performance of successful operations.

Assay of iron ore.—An assay of iron ore by fusion is not either a simple or an easy operation, especially if it is intended to separate, not merely the iron, but all the other ingredients; that is, make a quantitative assay. Iron ores are easily recognized in roasting, and there will be no need of smelting, if it is not desirable to know the quantity of iron contained in the ore. To smelt or assay iron ore, the ore must be brought to the highest state of oxidation; if it is not naturally a red oxide, it must be roasted to make it so. The ore must then be finely crushed, and the fluxes, also, pulverized. Five hundred grains, which is a little more than an ounce, which contains four hundred and eighty grains, form one charge. This quantity of ore must be mixed with one hundred grains of powdered limestone, one hundred grains of dried borax, and one hundred and fifty grains of hard charcoal. All these ingredients must be well pulverized, run through a fine silk sieve, and then mixed together. The common borax of the shops must first be dried — it would be still better if melted —which may be done in an iron vessel. If melted, it forms a clear and transparent glass; if it is only dried, it takes the form of a white spongy mass, having but little weight. Limestone is, for many reasons, preferable to burnt or slacked lime. If no borax can be obtained, a little common salt may be added to the lime, in addition to a small quantity of potash; but the results are not so certain as when borax is used. The addition of limestone is not always correct in principle, because some ores already contain lime; it answers for all the ores which have no lime in their composition. If there is any doubt as to the ore containing lime, it will be advisable to use only borax and carbon in the smelting, which succeeds in all cases; the only difference will be, that the assay will not be quite so correct without, as with the addition of lime.

One condition necessary for insuring success in an assay of iron is, to have good crucibles. Black-lead crucibles are, for various reasons, preferable to those made from clay; but, if the first cannot be obtained, the latter will answer very well, if the inner surface has received a good coating of black-lead, or charcoal, coke, or anthracite powder, moistened with clay-water, and well dried. The powdered materials for smelting should be well dried; if there is any apprehension of moisture, it will be advisable to dry tho ore and fluxes before the carbon is mixed with it. The crucible should be dry and warm, before the materials are put into it. The mixture of ore, flux, and coal should now be put into the crucible, pressed down gently, the remainder of the space filled with coarsely-pounded charcoal, and the whole covered by either a slab of clay, a piece of fire-brick, a large, hard piece of charcoal, coke, or, what would be preferable, a piece of anthracite coal. The materials should never more than half fill a crucible; the mass will boil as soon as melted, and may boil over the sides of the pot, if sufficient room is not left for the materials.

The crucible, well prepared, should now be placed in the furnace, upon a piece of fire-brick, or a broken crucible—the latter will answer best—so as to elevate the pot at least three inches above the bars of the grate. When the pot is firmly fixed, the fire should then be kindled, and kept at a brisk heat, with an open door, for at least one hour. After the lapse of an hour, the furnace must be once more filled with coal, which should cover the pot and its lid; the door should then be shut, and the most intense heat generated, which the fuel is able to produce. The bars of the grate should be repeatedly cleaned, so that the fire may receive a liberal supply of air. If charcoal is used as fuel, the fire must be replenished as the coal is consumed; coke and anthracite coal do not require additional fuel, if the furnace is once well filled. The latter coal requires a strong draft to generate heat enough for smelting iron. If the draft of a furnace is not sufficiently strong for the consumption of anthracite coal, charcoal, or a mixture of charcoal and anthracite, should be used. The second heat cannot be too strong, and should continue for at least three-fourths of an hour; after which, by the help of a pair of tongs, the crucible may be taken from the fire, and placed upon a heated brick. If the crucible should be placed, while heated, upon a cold, and particularly a damp place, it is liable to crack; in that way the whole assay may be lost. When the crucible has been slightly cooled, but is still of a white heat, it will be well to give it a few motions up and down, placing it gently but firmly upon its foot-piece. This motion is necessary, to settle the metal down, and collect together the particles which may be suspended in the melted slag. When the crucible has been so much cooled as to turn black, it may be quietly placed in cold water, and thus more rapidly chilled. When quite cold, the crucible may be broken, and the button of metal will be found in the bottom. If there has not been a strong enough heat, or the flux has not been sufficient, much of the iron may be mixed with the slag, which will contain it in round globules. The slag should be pounded with sufficient force to liberate the iron, without breaking its grains, and the pounded mass washed in the wash-pan, where the grains of metal will remain after the pounded slag has been washed off. An experiment in smelting, resulting as badly as the one above-described, would be considered a failure, not because the metal was not gathered together in one button, but because a large portion of the iron would still remain in the slag in the form of an oxide. In all such cases of imperfect smelting, the operation should be repeated. Want of heat may, in such cases, be a cause of failure; but, if the bulk of the metal has been accumulated into globules, want of heat was not the impediment to success—it was the want of a sufficiently liquid cinder which prevented the metal from running together. If limestone has been used in an unsuccessful experiment, it will be advisable to diminish the quantity of lime, and increase that of borax, in the next trial. A little common salt never produces any harm, and considerably facilitates the operation. Sometimes a failure is caused by the too free use of coal in the mixture, or by coal which has inadvertently fallen into the crucible: in these cases the cover of the crucible should be broken off, the coarse pieces of coal removed by a pair of tongs, and the crucible left uncovered, that the surplus coal may be burned off. The mass in the crucible will be found, in such instances, to be boiling; and the proper course will be, to increase the heat by cleaning the grate, and by the use of any other means that will effect that object. The smelting may then be continued until the boiling entirely ceases, after which the crucible may be removed from the fire, and treated as before described. We must be cautious in the use of potash or soda;

and it will be better if their use can be avoided. Borax and lime should be sufficient to finish any assay of iron.

The button of metal found in the bottom of the crucible, should be cleansed or freed from slag by the gentle use of a small hammer, and then weighed; the weight of this button is the proportion of metal contained in the ore. If 500 grains of ore are used, and the button weighs 100 grains, the ore contains one-fifth, or twenty per cent. of metal. After the button has been weighed, it may be broken, and the fresh fracture will indicate the quality of iron which may be produced from the ore, in large or extended operations. The knowledge required to enable an individual to judge of the quality of iron by the appearance of the fresh fracture, cannot be imparted in this treatise; it belongs to another department, and may be found in the author's "Treatise on the Manufacture of Iron."

Assay of copper ores.—If the ore is a conglomerate of native copper and rock, such as is found in large quantities at Lake Superior, and also in other localities, the operation is simple. The mineral should be coarsely pounded, and most of the rocky debris washed off; after this, it must be once more pounded, mixed with a small quantity of potash or soda, and exposed to a strong heat in a clay crucible, without any addition of coal. If the fire is strong, the copper button will soon be formed; fifteen minutes or half an hour is generally sufficient to smelt this kind of an assay. Copper ores, no matter of what kind they may be, whether poor or rich, sulphurets, or any other combination, are improved by roasting, which expels the larger portion, if not all the volatile matter contained in them. The roasting must be performed by a gentle heat, and the ore repeatedly pounded. The prepared ore is smelted along with black flux, or a mixture of black flux and borax.

Preparation of black flux.—Black flux is a reducing agent, and, at the same time, a flux for the rocky matter. It is a compound of carbonate of potash—pearlash — or carbonate of soda and carbon — charcoal,—and is made in the following manner. Two parts of crude tartar — also called argal or argol — and one part of saltpetre, should be finely pounded and mixed together, and then gradually heated in an iron pot so as to burn it. It should not be overheated, but, as soon as it is thoroughly warm, kindled, by means of a hot coal; it then burns slowly, and does not arrive at a heat sufficient to melt it. The mass, so prepared, is once more finely pulverized and sifted, and is then ready for use. It is very liable to absorb moisture, and should be protected against it by being placed in a tempered glass bottle. If tartar cannot be obtained, black flux may be made by dissolving sugar or starch in water, and mixing in this the potash or soda, in the proportion of one part of sugar to ten parts of soda or potash; they should be dissolved together and evaporated. The dry mass should then be pulverized, and treated as described. This last mixture works well, but is not equal to the first. If the latter flux is used, large crucibles will be required, it being liable to boil considerably.

Any of the copper ores, no matter of what description, if previously roasted, will make a good assay with black flux. The quantity of black flux used, varies between three and four parts of flux to one part of ore. If this mixture is not sufficiently fusible, the addition of a half or one part of calcined borax will furnish a liquid slag, which will permit the melted copper to pass to the bottom of the crucible. In this operation, a clay crucible, without either coal or plumbago, is preferred; too much coal being detrimental to the fluidity of the slag. In this assay, a quick heat is required; for, if the crucible be too long exposed to the heat, it is in danger of being cut through by the flux.

In addition to the above description, we will give the assay of copper ores, as performed at Swansea, in Wales.

Take one ounce of ore, pounded and sifted as fine as possible, calcine it in a pot, over a slow fire, until the blue flame disappears, which will take from twenty to twenty-five minutes. When this mass has cooled, it must be again pounded, and then mixed with the following flux, in the proportion of fourteen or fifteen pennyweights to one ounce of the ore. The reader will recollect that twenty pennyweights constitute one ounce.

1 oz.	8 dwts.	of nitre—saltpetre.
– "	8 "	of glass—common window glass.
– "	2 "	of calcined borax.
– "	2 "	of fluor spar.
1 "	12 "	of red argol—crude tartar.

This mixture must be pounded, and mixed with the ore in the proportion of three-fourth parts to one of the ore, put into a crucible, and melted. When the mass has become fluid, and does not boil any more, it may, to save the crucible, be poured into a hot iron mould or pot. As soon as it is cool enough to be solid, it may be immersed in water, the slag knocked off by the aid of a hammer, and the copper button weighed.

This assay does not always happen to be correct; the nature and composition of the ore occasion a necessary difference in its treatment. The quantity of flux given above is rather small, and is often increased to equal weights, that is, one ounce of the flux to one ounce of ore. Too much flux is considered to be as bad as too little; both retain copper in the slag, either in a metallic state, or as an oxide. A good button of metal has neither a very smooth nor rugged appearance; both indicate a loss of metal. It should have a faint copper colour, with a bluish cast, and not be too smooth. A whitish appearance of the button is decidedly bad, being the result of alloys which have combined with the copper, or of the flux, when it is in too great quantity: it may also be the result of too much argol, or of some coal which has found its way into the crucible. If there is any doubt as to the copper being all separated from the ore, the slag may be pounded, washed, and proper care taken to save it; if grains of copper are found in it, they may be added to the button. If there is still an impression that more copper is retained in the slag, in the form of oxide, which is indicated by its brown colour and porous texture, the slag may be finely pounded, mixed with a little quicklime, and melted again; the copper thus obtained may be added to the first. This last operation is always one of doubtful propriety; an assay which has been spoiled had better be repeated; it being an exceedingly difficult operation to extract a small quantity of metal from a red slag.

The copper buttons obtained in these assays are seldom or never very malleable, being generally hard and brittle. The appearance of these buttons, their colour, lustre, and texture, furnish indications of the correctness of the assay. Of these characteristics we shall speak, as they constitute the leading features of a copper assay.

If the button is very coarse, porous, and hard, and if it looks black, and does not resemble copper, the assay is lost; the yield or produce is not correct. A new assay must be made in this case. If the button is coarse, porous, and black, but slightly malleable, the assay may be good, and the yield true; but it will be necessary to re-melt this button, as we shall explain hereafter. Another instance is, when the button exhibits a shade of brown upon its surface; but this is, however, similar in other respects to those previously mentioned. The button yields, when re-melted, about two-thirds of its weight in fine copper. This button must be broken or pounded, and then melted or refined, with an equal or double weight of a flux, composed of crude or refined tartar, to which a little common salt has been added; to prevent the appearance of a red or burnt slag, a little charcoal or coke, or some powdered stone-coal may be also used. The next species of button has a brownish-red or copper colour; this may be considered a good result, and giving a true yield. If this button should be broken, and melted with twice or thrice its weight of the above-mentioned flux, it will yield one-half its weight in fine copper. The worst description of buttons are those having a white, grey, or yellowish colour; these are useless, and the assay is a complete failure.

The first operation in these assays is, the formation of crude copper, which must be refined by a second process of smelting. In this second operation, as much flux must be constantly present, as will properly cover the metal, and prevent its oxidation and consequent loss of copper. For the second smelting or refining, the best flux that can be used is the black flux, which we have previously described, with the addition of a little common salt.

If, in the first assay, the ore does not contain sufficient iron, which will be denoted by the dryness and porous character of the slag, and the coarse and porous nature of the button, some iron must be added to it. An addition of flux may assist in remedying the evil, but there is a danger of the retention of copper in the superfluous flux. Iron is, in such cases, the best flux; it must be used in the form of a fine powder, and as an oxide, or iron ore. There is a method of directly melting such dry ores, and obtaining from them a good button; but it is not quite so certain to produce a correct yield. This operation will be described in the succeeding pages.

To produce a good button at the first melting, the flux necessary for the performance of the operation must be compounded as follows:—

3 oz. of argol—crude or refined tartar.
1 " 3 dwts. of nitre—saltpetre.
– " 10 " of calcined borax.
– " 10 " of bottle or window glass.
– " 6 " of slacked lime.

These ingredients should be well mixed together, pounded, and then sifted; the flux will then be ready for use, and must be kept in a tightly-closed bottle, to protect it from the moisture.

One ounce of calcined ore, prepared as described above, should be mixed with one and one-half ounces, or a little less, of the last-mentioned flux. About two or three pennyweights of common salt, and an equal quantity of argol should be pounded and mixed together, put into a small crucible, and melted, at the same time that the ore and flux are being melted in another crucible. When the latter is melted, and ready to be withdrawn from the fire, it should be emptied into the other crucible, in which the flux is also in a fluid state. The motion produced by the discharge of matter from one crucible into the other, will have the effect of gathering together all the stray grains of copper, which might, otherwise, have adhered to the sides of the crucible. When the latter crucible has been heated to the proper degree, the contents should be poured into an iron mould, previously heated and greased; or, it may be allowed to cool, and the crucible be broken, to obtain the button of copper. The copper thus obtained, is frequently found to be crude, in which case it will be necessary to refine it.

This last mode of assaying is never quite correct, and always falls short of the true yield. It is a profitable assay for the purchaser of ore, but an unprofitable one for the seller.

Rich ores are very sensibly affected by fluxes, and, consequently, more subject to a loss of metal than poor ores. To prevent this, the amount of saltpetre and salt may be diminished, or entirely dispensed with. Such ores are frequently melted with argol alone, to which a small quantity of coal-dust has been added. After all, such ores never produce a good yield when assayed.

The surest and best mode of assaying copper ores is, always to divide it into two operations. If it can be so arranged, and performed in such a manner as to lose but twenty-five per cent., or one-fourth, in refining the button, the operation may be considered to have succeeded very well. If the refining wastes more than one-third of the crude copper, the first operation must have been imperfect, and the assay may be said to be lost. A little experience will teach any one whether the ore melts too sluggishly or too rapidly, on the application of the first heat. If it melts too rapidly, the flux is too sharp and too fluid, and the assay is spoiled; but, if it is too tough in the crucible, the addition of a little calcined borax, or fluor spar, will make it lively: the use of too great a quantity of these articles, has, however, a tendency to make the copper crude.

Refining.—The copper obtained from these assays has not yet acquired a fine quality; it is more or less hard and brittle. The buttons are, therefore, once more broken, to prepare them for the process of refining. For this purpose, a crucible is placed in the furnace, and brought to a white heat; the crude copper having been heated in a small crucible, must be thrown into the hot one, and melted rapidly. As soon as the copper has melted, a little of the following flux, which has been previously melted in another crucible, must be poured over the hot copper, and a few minutes after, the contents of the crucible should be emptied into a greased mould. If the copper does not prove to be fine, after the last melting, the operation of refining will have to be repeated, in the same crucible, until the copper has acquired the necessary fineness. Copper is tested by flattening it on an anvil by means of a heavy hammer, bending, and then breaking it. If colour and strength are satisfactory, the assay is, thus far, finished. The slags resulting from these refining operations, should be mixed with a small quantity of argol—tartar,—and smelted in the same crucible in which the refining was performed. The grain of copper resulting from this last smelting, should be added to the weight of the button of fine copper, and will complete the assay. This grain often has the appearance of iron, and should be refined. After one or two smeltings, nothing will remain but pure copper.

A refining flux for coarse or crude copper:—

1½ lbs. of nitre,
1 " of argol,
½ " of common salt.

Mix and pulverize these materials well together, and put them into an iron pot; then heat a poker, or some other thin piece of iron, to a red heat, and stir the mixture with it, until it is properly and thoroughly burned. After this, pulverize it, and bottle it for use.

Another flux, qualified for use in copper of a more crude character than the one previously mentioned is suited to, may be compounded of the following ingredients:—

2 lbs. of saltpetre,
1 " of tartar,
½ " of common salt.

Assay of lead ores.—An assay of lead ores is not very difficult, but it is still an operation requiring more skill than one would, at first sight, suppose. Lead is so very volatile, that, if the smelting operation be not conducted with great care, considerable losses will occur, rendering the result of an assay very doubtful. These losses by evaporation amount, in many instances to twenty, and sometimes, even to thirty per cent.

For the sake of convenience we will divide the lead ores into two classes: one class will comprise those ores containing no sulphur; the other class may be mixed with sulphur, or be pure sulphurets. The first class can be assayed without the addition of iron; the second requires the use of metallic iron to produce satisfactory results.

Lead does not require much heat, and a high heat should not be used, if losses are to be avoided. A common stove, a fire-grate, or any description of furnace, will suffice for this purpose. Graphite crucibles answer the purpose better in this assay than those composed of pure clay; if the latter are used, they should be lined with a coating of charcoal powder or plumbago. An assay of lead can be performed to the greatest advantage in a cast-iron crucible. Instances occur, in which a great heat is required to smelt a lead ore; but these are extreme cases, and should never be depended upon for correct results.

Lead ores of the first class, or those which do not contain any sulphuret, should be smelted simply with black flux. If the lead ores are very pure, such as litharge, minium, white lead, and similar ores, there will be no flux, or very little of it needed. Common lead ores, such as are in a natural state, are never sufficiently pure to yield their lead when smelted, without the addition of flux. Such ores should be pounded, and roasted gently, or merely dried, to expel all the water and vegetable matter which may be contained in the ore. The powdered ore may be mixed with pulverized black flux, in the proportion of one ounce of ore to two ounces and more of the flux. Two parts of flux to one of the ore is generally sufficient; but there are cases in which that amount will not even make a fusible slag, and more flux will be necessarily required. There is no objection to using as much flux as we please, but then it compels us to use larger crucibles, as this flux is very liable to boil. In cases in which two parts of black flux are not sufficient, it is advisable not to use any more flux, but to add carbonate of soda, alone, or in connection with a little glass of borax. With these additions to the black flux, the process is always successful; provided we use a plumbago crucible, or one lined with charcoal or plumbago.

The mode of performing the operation is very simple: when the flux and ore are well mixed, it should be put in a crucible, previously heated. The crucible should not be more than one-third full; the other two-thirds of the room being needed for the ebullition. The crucible should then be covered by a slab or a good hard coal, and exposed to the heat. The crucible should be placed in a cold furnace, the fire kindled, and the heat raised gradually; as a sudden or rapidly-increasing heat is very apt to boil the mass violently, and, if the crucible be not very large, drive the contents into the fire. A layer of common salt over the top of the ore will moderate the ebullition; but if this does not prove to be sufficient, and there is any danger of the mass boiling over the top of the crucible, the cover should be taken off, and air admitted, which will speedily diminish the agitation. As soon as the boiling ceases, a stronger heat may be generated, and the mass brought to a state of perfect fluidity; but a greater amount of heat should never be applied, than will be actually necessary for the accomplishment of our design.

When the mass is properly melted, and appears in the crucible with a clear and glassy surface, the assay is accomplished; the crucible should then be taken from the fire, cooled, quenched in water, and finally broken, to remove the button of lead, which is found in the bottom of it. There is generally no difficulty in separating the button from the slag, but it sometimes adheres to the crucible, when made of clay or iron. When this occurs, a great deal of chiselling is frequently necessary to separate the adherent particles. It is necessary in every case to pound the slag, wash it, and examine whether there are any grains of lead in it. The slag being usually soluble in water, nothing is needed but to pound it coarsely and dissolve it; if there are any metallic grains contained in it, they will remain after the slag has been washed off. If the mass has boiled too strongly, and the fire has been too hot, the crucible will be very apt to retain some of the crude slag, and form globules of lead, which will adhere to the sides of the crucible. The loss of these globules diminishes the yield of the ore, and makes the assay incorrect. Such an assay cannot be remedied, and is lost; the adherence of these globules to the crucible having been caused by the boiling, every effort should be made to prevent that inconvenience.

The lead produced by such an assay is never pure, and always retains some traces of those metals which have been mixed with the ore, such as copper, silver, antimony, &c. It also invariably contains some potassium or sodium, which it receives from the fluxes. These admixtures are, however, in such limited quantities, as to exercise but little influence upon the correctness of the assay. The loss by evaporation is of far more consequence, amounting generally to about ten or twenty per cent. If zinc and lead are found united in the same ore, and the proportion of the former be greater than that of the latter, the lead will frequently evaporate entirely. In this case the only chance of success exists in washing the ore, roasting it, and using an excess of flux in smelting it, taking care to use only as much of the black flux as will barely precipitate the lead.

Lead ores of the second class, or those which contain sulphur, should not be roasted, but simply smelted with black flux and a little dry carbonate of soda, with the addition of some metallic iron. Twenty per cent. of the latter, in the form of wire ends, cut into short pieces, tacks, or broken nails, is generally added; but no harm will result from the use of a larger quantity. Half the amount, or even an equal weight of iron to the ore, will have no unfavourable effect, if the heat in smelting be not so great as to reduce or melt some of the iron. Iron of a coarse form is the best for this purpose, as it can be used liberally without any resulting evil effect. The quantity of flux may be varied, without occasioning much difference in the result, from one-half to double the quantity of the ore. If the ore is a simple sulphuret, like galena, carbonate of soda and metallic iron will be sufficient for smelting it; but if it contains any oxide, or other forms of lead, the black flux must be applied.

The most perfect, and, at the same time, the surest way of assaying lead ores is, in an iron pot or crucible. Any iron pot will answer this purpose, provided that it be not too large. If one can be obtained, having the form of a clay crucible, it will be well to procure it; still, any pot having a round bottom, is good enough for an assay of lead. If black flux, soda, or nitre, cannot be obtained, a very efficient flux can be compounded of potash, to which soap has been added to an amount equal to the weight of the ore, and a small proportion of common salt. This flux should be mixed with the pounded ore, and the mixture dried in an iron pot over a gentle fire; care being taken to stir it constantly to prevent its boiling over. When it is thoroughly dry, the stirring should cease, and the heat be gradually raised to the melting point. Sulphurets, in particular, work well in an iron pot. The best assay can be made in an iron crucible; it furnishes the purest lead and the largest yield: if the loss in a clay crucible will reach twenty per cent., it will not average more than ten per cent. in an iron crucible. When iron pots are used, the addition of metallic iron will be the same in quantity as before designated. All assays of lead do not furnish a full yield.

Assay of gold ore.—Gold is frequently, and, we may say, usually found in its native condition. It is also found, mixed with metallic sulphurets. The latter ore may be considered the most abundant in the United States. As gold is one of the most extensively distributed minerals in this country, we shall devote to it more attention than we should have been, otherwise, inclined to do.

If the gold is in its native state, enclosed in other minerals, the ore may be pounded and washed, and the gold separated from it and weighed; but such an assay will be incorrect, as a large proportion of the fine gold will be washed away, and more than half of the actual contents of the ore lost. Such an ore may be amalgamated; but that process is, also, not quite safe, and remarkably slow and tedious. The best way is, in all cases, to make an assay by the process of smelting.

Gold ores are generally poor; it is useless, therefore, to smelt as small a quantity as one ounce, five ounces being little enough in any case for the production of a successful assay. The ore should be finely pounded, sifted, and at least an equal weight of litharge added to it; in the plurality of cases, twice or thrice its weight of litharge may be required. When the ore is very siliceous or sulphurous, this will be the case; in fact, it will be best to use always a liberal amount of litharge, as it never does any harm. The litharge sold in the shops is not always pure, sometimes containing copper and iron, and frequently silver, which is worse than either, rendering an assay made with it incorrect. The surest method of obtaining pure litharge is, to take acetate of lead, and dry it by a strong heat, constantly stirring it, at the same time, to prevent it from melting. The litharge obtained by this process may be considered free from silver. To the mixture of ore and litharge add a little black flux, just sufficient to afford carbon enough for the precipitation of a limited quantity of metallic lead. One part of black flux will produce one part of metallic lead; and as half an ounce of lead will be sufficient to absorb all the gold contained in five ounces of ore, not more than half an ounce of black flux should be added. If there are no pyrites in the ore, some carbonate of soda or borax may be added; but if sulphurets are present, it will be advisable to use nitre. In most cases, the assay will be more perfect, if we abstain from the use of alkaline fluxes entirely, and substitute litharge and a small quantity of finely-powdered charcoal. The assay may be performed in a crucible composed of pure clay, or, which would be preferable, in an iron pot. When smelted in the latter, the contents should be emptied into a mould while hot, because it will be difficult to separate the mass from the pot, when it becomes cold. The metal in the pot should be invariably covered with a layer of salt, as it facilitates the smelting, and the separation of the gold from the slag.

The lead obtained in this instance, contains all the gold derivable from the ore; it must be cupelled, an operation which we shall describe hereafter. One of the principal conditions of success is, an accurate proportionment of the ore and the flux, as well as of the carbon. For this reason, black flux is preferable to charcoal, but soap is better than either; soap penetrates every ramification of the mass, and its carbon is brought into close contact with every particle of the ore. The application of alkaline fluxes is objectionable, so far as it causes an ebullition of the mass;

and if sulphur be present, it will retain a portion of the gold.

Assay of silver ores.—Native silver is seldom found in a sufficiently pure state to afford a chemically pure silver; on this account it should be smelted with a little metallic lead, as this will absorb the impurities which may be enclosed in the mineral silver. Native silver is usually found alloyed with gold, which may be separated by the dry analysis; but the humid method is much the most simple and certain in its results, as will be demonstrated at the conclusion of this article.

Silver ores are found in a great variety of forms; the metal combining with almost any kind of material, to form an ore. If the amount of silver contained in a mineral is too small to repay the cost of extracting it, it is not regarded as a silver ore, but is used for some other purpose. Silver ores are assayed very much in the same manner as gold ores, differing, however, in the degree of heat used; silver requiring a higher temperature to smelt it than gold does. Most of the silver ores contain sufficient lead to make the addition of that metal superfluous; in this case the ore should be treated exactly as a lead ore, with the exception, that little or no iron must be used in the assay, but principally black flux and borax. It is a matter of but little consequence if lead remains in the cinder, for the greater portion, if not all of the silver, will follow the bulk of the lead. All the silver will be obtained by the application of a strong heat, no matter how much lead may accompany it. The desideratum in this assay is, to produce but little lead, and obtain all the silver; this can be accomplished if the ore is properly roasted—often a tedious operation,—and the assay be made with saltpetre and argol, to which borax may be added. If the assay be performed in this manner, it will be perfectly safe, provided all the conditions be complied with; but as no reliance can be placed upon the adherence to those requisites, it will be advisable to follow the subjoined rule, which, though more tedious, is certain of success. If the ores are rich in silver, a portion of litharge, made of acetate of lead—sugar of lead—should be added; the quantity of litharge may be optional, but should always be in an inverse proportion to the amount of lead contained in the ore. It is not necessary to have so large a quantity of silver ore as would be required of gold ore, in these operations; one ounce may be considered in all cases a sufficient quantity for obtaining practical results. A very perfect solution of iron is needed in the flux, and, if the ore contains much silex, lime, or other impurities, which cannot be removed by washing, a large quantity of litharge will be required.

When the ore has been washed and finely pounded, it should be mixed, as the case may be, with more or less litharge—it will not, in some cases, require any. An addition of half an ounce of black flux should be made, which will produce half an ounce of lead. The powder of charcoal made of hard wood, is better than black flux; one part will make thirty parts of lead, reduced from the oxide. Eight grains of charcoal will make 240 grains, or half an ounce of lead. In every instance, a half ounce or one ounce of dried borax should be added to the mass, which should be thrown loosely into a dry crucible, and not more than one-third fill it. A layer of common salt should be spread over the ore, the crucible covered, and then exposed to heat. The heat should be strong and rapid, and consequently the application of alkaline fluxes would be injudicious; litharge and glass of borax should finish such an assay. Boiling should be avoided in this case more than in any other; if it cannot be entirely prevented, it should be diminished as much as possible.

If the silver ore is a perfect sulphuret, that is, if all the metals in it are sulphurets, the assay may be made comparatively easy. When this is the case, the crude ore should be finely pulverized, and mixed with an equal weight of litharge, and nearly half its weight of nitre. If there is a large proportion of iron or copper pyrites in the ore, the amount of litharge and saltpetre should be proportionably increased. Iron pyrites require four times the above weight of nitre, copper, three times that amount. This mixture will produce all the silver in but a small quantity of lead. If the alloy contains too much saltpetre, it will not produce any lead at all; but this evil may be remedied by adding some small chips of metallic lead to the fluid mass. Any ore will readily yield its silver on the liberal addition of litharge; while the presence of alkaline fluxes is not so favourable to the precipitation of silver, but has a tendency to produce more lead.

Refining lead, to obtain the gold and silver.—The silver and gold in the foregoing described assays, having been alloyed with a large portion of lead, the following assay by the cupel, is designed to destroy the lead and other metals, and produce the gold and silver in their pure state. This process is founded on the feeble affinity which gold and silver have, comparatively with other metals, for oxygen. The oxides formed in this assay, melt and sink, like a fluid glass, into the pores of the cupel, which are not, however, large enough to absorb the metal.

Cupel.—We have previously alluded to the formation of a cupel for small assays, similar to those we have been describing; but, as cupelling is a delicate operation, and, as success chiefly depends on the cupel, and, as this operation must be frequently performed in smelting establishments, we will, in addition to our former description, allude to the manner of making large cupels for practical use. Take bones and burn them well in an open fire, so that they will be thoroughly white when broken. After this, pound the bones well, and sift the powder through the finest wire sieve which can be obtained, or, which will be still better, through a sieve made of bolting-cloth. Put the pulverized bone ashes in a wooden tub, and pour on boiling water; stir the mass well and let it settle: as soon as it has settled, pour off the clear, supernatant water from the sediment, and repeat the process, which should be done two or three times. The washed bone ashes should now be dried, and again moistened with a little water or stale beer, but not so much so as to make the mass sensibly wet. The damp ashes should then be filled into an iron ring, three or four inches in diameter, and one inch high, and rammed down as tight as possible, flush on both sides of the ring. One side should be then scraped out, so as to form the cavity for the reception of the lead to be refined, and then made perfectly smooth. It should be dried in a gentle heat, to expel all the moisture, which will require at least two days. This cupel will hold from three to six ounces of lead, and, being large, will need an iron hoop, to guard against the danger of breakage. The amount of moisture to be used is a nice point, and requires close attention. If there is too much water in it, the mass will shrink, form cracks, and perhaps fall out; it will, at all events, be unfit for use, for the cracks will absorb the metal, and spoil the assay. If there is not sufficient moisture in the ashes, they will not adhere together, and will consequently fall out. As much moisture as is necessary to hold the ashes together is all that is needed, and makes the best cupels. To make the best work, the ashes should be well kneaded, and strongly squeezed. A cupel will absorb an equal, or nearly equal weight of lead, that is, a cupel of one ounce weight will absorb one ounce of lead.

Muffle.—This is a box made of fire-clay, shaped like a travelling-trunk, in which one head is wanting. The clay must be from one-fourth to three-eighths of an inch thick; it is usually from ten to twelve inches long, six inches broad, and four inches high, and is provided at the top with numerous small, oblong holes. The muffle should be enclosed in a furnace by a wall, in such a manner, that the fuel may completely surround it. It should be composed of good fire-clay, capable of resisting a strong fire. In this muffle the cupels should be placed for refining. They may be purchased ready-made in the large cities, but, at a distance from such places, it is difficult to obtain them. Instances may occur, rendering an assay imperatively necessary; and, if no muffle can be procured, a good-sized crucible may be selected, a hole driven through the bottom of it, and the cupel containing the lead placed in it. The crucible may then be placed in a furnace, in a position that will admit a draft of hot air through the bottom of it. When a reverberatory furnace is at our disposal, the cupel containing the lead may be placed directly upon the fire-bridge: this may also be done in the furnace of a steam-boiler, if the bridge is hot enough. The crucible is, however, the safest substitute for the muffle.

The process of cupelling is very simple; the muffle should be heated, and the cupel, containing the lead to be refined, placed in the mouth of it, so that it may receive the heat very slowly. When all the moisture has been expelled from the cupel, it may be pushed farther into the muffle, and a stronger heat applied. In a short time, the bone ashes will have a very white appearance, but the metal will look red; if there is an iron hoop around the cupel, the metal will assume a dark red colour. Many operators prefer leaving the addition of the test-lead until the metal has been thus far heated, fearing some accident might happen to the cupel, and the lead be lost. The heat should now be rapidly increased, but not made too intense. If the lead becomes too lively, and particularly if there is a considerable evaporation, the heat should be reduced, or the vapours of the lead will carry with them the silver and gold. The best heat is that which is just sufficient to prevent the lead from chilling, and heat the oxide of lead to the degree necessary to allow it to sink in the cupel, and not form a cold black ring around the hot lead: as soon as this happens, the heat should be increased, and the cupel will absorb all the oxide made by the test. The lead should not only be kept liquid, but also in lively motion, until nearly the whole of it is absorbed by the cupel; the heat should then be rapidly increased to the highest degree—a bright white—to melt the scattered silver or gold into a round globule. When all the lead has disappeared, and not previously, a bright, shining, metal globule will appear. The cupel should now be removed from the fire, and cooled; when cool the globule may be separated from it, cleansed from the adherent dirt, and weighed. An assay of this description is never quite correct, the yield being always too small. In large operations, the yield will be greater from the same ore or lead. If the difference of yield in large and small operations were always the same, there would be no evil resulting from it; but this is not the case; the multiplied and complicated operations necessary in assaying, cannot be conducted with so much precision as to produce uniform results. All assays which do not work well, can never be depended on.

In all assays made immediately from the cre, silver will be found more or less alloyed with gold, and gold with silver. As the separation of these two metals by the dry analysis is uncertain and tedious, we will describe a process by which an approximate result may be obtained. The proper method of separating gold and silver is, by the humid process; that is, dissolve the alloy in nitric acid, and precipitate the silver by means of a solution of common salt. The explanation of the humid process belongs properly to the department of chemistry, and is, therefore, beyond our limits.

To separate gold from silver and other metals by the dry analysis, it should be mixed and melted with three times its weight of crude antimony — not the metal, but the sulphuret of antimony, obtainable in the drug stores. This alloy should be run into a mould, cooled, the slag knocked off, and the brittle regulus or button melted in a fresh crucible, and treated as before. In the second smelting, some saltpetre and a little common salt should be added; these will destroy and oxidize all metals except gold. If, after the pot has been cooled and broken, the button should be found brittle and hard, the same smelting process, with the addition of saltpetre and salt, should be once more repeated. If, after repeated smelting, the gold still remains impure, more antimony should be melted with the gold, and the refining process repeated, until the experiment is successful. By this method all the gold will be obtained pure; and the difference in weight between the gold, and the metal received from the cupel, will be silver, if cupelled metal has been used. All the metal to be assayed by this operation should be the result of cupelling; and, for this reason, all gold or silver ore from which it is desirable to obtain pure gold, should be melted with lead, and cupelled.

If the gold is in fine grains, such as wash-gold or amalgam, it should be purified by mixing it with a little corrosive sublimate or calomel, heating it at first gently, and then raising the heat until it melts the gold. This operation should be performed in a new crucible; the silver, and other metals, are then lost by evaporation.

If, in the foregoing operations, any silver should remain in the slag, the silver may be recovered from it, by smelting it with argol, to which a little charcoal dust has been added. The recovery of the silver will, however, be more certain, if the slag should be treated as a poor silver ore — pounded, smelted with litharge, and then cupelled, as previously described.

If gold or silver is alloyed with platinum, iridium, rhodium, and copper, the separation of these metals will be very difficult; the alloy must be subjected to the humid analysis. In all such instances, it is utterly useless to endeavour to do good work by the dry assay. Such alloys may be melted with lead; but, in cupelling, they cannot be separated,—not even the copper will, in this case, leave the other metals.

Assay of the ore of mercury. — Cinnabar is the common quicksilver ore, and we have but this one kind to operate upon. In performing the assay, it makes but little difference whether or not any other description of ore be present. The cinnabar should be pulverized, and mixed with half its weight of iron filings or borings, and some slacked lime or soda, the mixture put into an iron retort, and the retort exposed to a strong red heat. The heat should be raised slowly, as it will otherwise be likely to break the retort. The neck of the retort should be prolonged by an iron pipe, three-fourths of an inch wide, and about two feet long, which is surrounded by another pipe, made of tin, sufficiently large to leave a space between the two; into this pipe a current of cold water may be admitted, which will flow in at the lower end, and pass out at the upper or highest part of the tin pipe. The current of water will keep the iron pipe cool; this condenses the quicksilver in the iron pipe, and it will pass from it in a metallic state in drops, which may be collected in an iron or porce lain basin, filled with water. A strong heat, and a sufficient amount of iron, will drive the last traces of mercury from the ore. The fire should completely envelope the retort, to prevent the adhesion of quicksilver at the top. If the iron pipe be smooth on the inside, very little mercury will adhere to it. A retort should never be more than one-third full; the ore being very expansive, will burst the retort if it be too full.

Instead of an iron retort, a stone-ware bottle, or an earthenware jug may be used; care must be taken to prevent the heat from breaking the vessel, as this description of ware will not bear much change of temperature. If an iron retort, of the above description, cannot be obtained, it will be better to use a common cast-iron coffee-kettle, which can be procured in almost every place. The lid of the kettle may be cemented tightly upon it with strong clay, moistened with salt water, or a solution of glauber salt, and be held down by an iron bar, passed through both of the ears, to which the handle is attached. A glass tube may be used for a pipe, but an iron one is preferable. If a wrought-iron gas pipe cannot be procured, a substitute may be made of thin sheet-iron, the joint well closed, and the pipe bound with iron wire. The joint should be turned upward, and made tight by the application of fine clay. The end of the pipe should be led over a basin containing water, so that the mouth will be about two inches above the water. A moist rag may be hung over the end of the pipe, and kept constantly wet, so that the vapours of the mercury, which descend the pipe, may be condensed in the rag, and drop into the basin of cold water.

The quicksilver gathered in the basin, should be put in a bottle and weighed; the weight indicates the yield of the ore. Mercury obtained in this way is not always pure; but, as the result of an assay, it is sufficiently pure to be depended upon; the amount of impurities never being very large.

Assay of tin ore.—The assay of this ore is one of the most difficult of the whole series. Tin is so easily oxidized, and the oxide combines so readily with siliceous matter, that it is almost impossible to effect a perfect separation of the metal from its ore. Tin ore should be well washed and purified, before it is exposed to heat in the process of an assay; tin ore being very heavy, and easily separated from other matter, this may be readily accomplished. A practical method of assaying tin ore is followed at Swansea, in Wales; and, as mathematical precision is out of the question in this assay, we may just as well depend upon the assay as it is performed there. Take five ounces of clean ore, either black tin, that is, oxide of tin, or any other tin compound; mix the powder of the ore with one-fourth of its weight of culm, or coarsely-powdered bituminous coal; and put the mixture into a warm, or dry blue pot—black-lead crucible. The pot should then be exposed to a strong and quick heat — the most intense that can be produced — for a period of fifteen or twenty minutes; or, until the mass has ceased boiling, and has quietly settled down in the pot. When perfectly fluid, it should be stirred with an iron rod, and receive another heating, for the space of five or six minutes. The crucible should then be removed from the fire, the tough slag on the surface of the metal pushed aside, and the contents poured into a metal mould. The slag should be scraped out of the crucible as well as possible, into a basin, and carefully saved. The metal in the mould must be cleared of the slag adhering to it, all the slag thrown together, pounded, and washed; by this operation many grains of tin may be found, and added to the ingot. In this way, the ore will yield a great amount of tin, but not all that it contains. The process used in the tin works, for testing the ores, and deciding their value, answers the buyer very well, but is not so profitable to the vender of the ore.

If a new crucible be used in the assay, the yield will be always from eight to twelve per cent. less than from an old crucible. The safest assay is, a second one, performed in the same crucible; it will always be found more correct than the first. The metallic tin made by this assay is never pure; it often contains twenty per cent. and upward, of other metals, which must be subtracted from the actual yield of the ore. In pure ores, this loss is found to be but eight or ten per cent. If culm, or powdered bituminous coal, cannot be obtained, dry sawdust may be used; charcoal does not work well, and may be substituted by brown wood or brands.

A correct assay of tin ore can only be made by the humid process; or, as some prefer, by a partially moist and dry operation. The safest method of ascertaining the exact amount of tin in any ore is, the chemical analysis, made by an experienced chemist.

Assay of zinc ore.—Zinc is too volatile to have its ores reduced to metal in the usual way. The distillation of zinc requires a good retort of iron or fireclay, either of which is expensive; and, after all, an assay of zinc, made by distillation, on a small scale, is incorrect. The best plan for the assaying of zinc ore, in case it is blende, or sulphuret of zinc, is, to roast the ore well, along with some powdered charcoal. Oxides of zinc do not require roasting, but they may be heated sufficiently to drive off the water and carbonic acid which may be contained in them. The ore, thus prepared, should be once more finely powdered, mixed with fifteen or twenty per cent. of charcoal powder, and exposed in a plumbago crucible to a strong heat. The ore should be perfectly dry, closely packed in the pot, and covered by a slab or an inverted crucible. A strong and rapid heat will evaporate all the zinc, and the remaining matter may be scraped out of the crucible, pounded, and exposed to a red heat, in an open dish, to burn off the carbon contained in the mixture. The residue must be weighed after this last roasting, and the loss which is found to exist between the original weight of the ore, and the residue, will be oxide of zinc, from which the metallic zinc may be deducted: 100 parts of the oxide are equal to 81 of metal. The assay made in this manner is perfectly safe, provided the ore contains but little or nothing of other matter; if it contains lead, tin, silver, gold, copper, antimony, and similar metals, they will all be carried off by the vapours of the zinc. Some of these metals will be lost by the evaporation of zinc in every instance, no matter how small the quantity may be which the ore contains. Zinc ores which are not intended for use in the manufacture of zinc, cannot be assayed in this manner. If it be intended to apply the ore to the manufacture of zinc, it will make no difference what number or quantity of foreign metals is evaporated.

Assay of chrome ore.—We should not allude to the assay of chrome ore, were it not for the circumstance, that it is extensively found in the United States. If we merely wish to ascertain whether there is any chrome contained in a mineral, it will be sufficient to pound it, mix it with its weight of potash or nitre, or both together, and smelt it. If, when almost cold, the crucible be immersed in pure water, the potash will dissolve in it; the smallest amount of chrome will colour the solution yellow, if too much water be not used. If there is any manganese in the ore, the solution will be green; by exposure to the air in a flat dish, the manganese will shortly subside, and the liquid will assume the yellow colour. A quantitative assay of chrome ore is beyond our limits, and belongs to analytical chemistry.

Assays of the ores of cobalt, nickel, antimony, bismuth, and other minerals of similar composition, are beyond our province, and must be submitted to professional chemists.

PART III.

MINES.

MINES are the subterranean treasuries of Nature, in which the omniscient Being has deposited a portion of His bounties, to be developed and used by intelligent men. Mining, improved to its utmost extent, is an art, in which all the lights of science, all the capacity of mind and diligence of man, find their application. In the United States the mining operations are of too recent an origin, to present any masterpieces in the art. It requires ages, the lapse of centuries, to develope such subterranean structures as can be found in the Old World, and, also, in our sister Republic of Mexico. An extensive subterranean mine is a grand and curious structure, and possesses many beauties which are understood only by the initiated. These cannot be made the subject of panoramic views, like the terrestrial beauties of nature, and are, therefore, not accessible to the multitude as a means of enjoyment. In subterranean mines, there is poetry, a high, religious poetry; the miner, in his lonely chambers, is constantly reminded of the bounties at his disposal, and the assistance he needs from a higher power than what man can bestow. We regret exceedingly that it is not permitted to us to luxuriate in the description of mining to its fullest extent; our aim is to be useful, and as there can be no immediate use in describing the highest cultivation of this art, it would be wasting the time of the reader, to extend this essay farther than its practical utility will warrant.

Mineral deposits cannot be discovered, except by actual search. There are leading features pointed out by geology, but these can do but little more than give us an approximate idea of the locality where minerals may be found. Geology furnishes us with instructions sufficient to enable us to determine what kind of minerals may be found in a certain description of rock. Besides geology, the assistance of geometry will also be required, to point out to the miner the place where he may find mineral deposits located. Among the fallacious indications of minerals, are those derived from unimportant circumstances, such as the issuing of mineral springs, the emission of vapours from crevices in the rocks, the more rapid melting of snow in one place than in another, and the presence of certain species or kinds of vegetation. The divining-rod, and the compass made of load-stone or iron pyrites, are also an absurd piece of folly. Such means tend to support the pretensions of the deceiver and impostor; and, it is to be regretted, have often been the cause of superstitious and ignorant persons engaging in researches, which were absurd in their nature, and often proved ruinous to the credulous treasure-hunter.

Indications of a deposit of minerals are either indirect or direct. Indirect indications are those pointed out by geology. We cannot find water by digging into drift sand, unless we dig through it; bituminous coal cannot be found in granite, nor gold in the coal formations; tin is not discovered in limestone, nor spathic iron ore in secondary rocks. If such things do happen, it is merely an exception to a general rule. Positive or direct indications of the presence of minerals are, the finding of a specimen, even though not in the proper place; the frequent occurrence of fragments of minerals strewed over the surface; and the actual discovery of a vein or deposit.

HUNTING FOR MINERALS.

In hunting for ores or minerals, the first thing necessary to be ascertained is, whether there be any mineral in that locality, and of what description. The next question will be as to the precise spot in which it will be most likely to be found. Suppose we wish to find a coal vein which we know to be in the vicinity, we must, in order to insure success, know the accompanying rock, that is, the rock upon which the coal is imbedded, as well as that which rests upon the coal. Knowledge of this character multiplies the facilities for finding the coal. If the coal be three feet thick, and we know twenty feet of rock below, and twenty feet of rock above, this facilitates our labours, thirteen to one. In such cases, we select the rock accompanying the mineral, or such a vein of dead mineral as has an extensive range, no matter how far, how high or low it may be from the vein. We measure the distance between both; and, if the rock wherein the mineral vein is buried belongs to the stratified rocks, we may expect to find the vein at no great distance, in any direction. If we know the vein-stone, or accompanying rock, we may rely upon finding the mineral, even though we do not discover it at the place where we open the vein, or touch the vein-stone. Where the veins run parallel with the strata of rock, as is the case in the coal region, and also in the gold region of the Southern States, there is no difficulty in finding a vein; but where the mineral veins traverse the strata, in angles more or less defined, the object is not so readily attained. In the latter case, the general direction of the vein is marked by the magnetic compass, and followed by its indications. This mode is sufficient, where a heavy body of iron does not interfere, and cause the needle to vary from its magnetic meridian. Operations of this kind must be performed by a scientific miner.

When fragments of ore are found on a hill-side, it is very evident that the vein must have a higher location, for these fragments very probably came down the hill. If the position of the vein is horizontal, and fragments of its minerals are found on the top of the hill, there is no probability of finding the vein, for it is generally entirely washed away. If fragments of minerals are found in a stream, the veins which supplied them must be higher up the stream; and the farther up the lighter the material. Heavy minerals do not drift far, and, in consequence of their weight, are easily destroyed. Gold never drifts far in a stream; it is always found close to its source. If the current of water is strong enough to move grains of gold, it soon rubs them into such a fine dust, that it can be carried off by the most gentle current. Therefore, gold is seldom found in the beds of rivers. Native metals, and sulphurets of metals, are always found near their source, because they cannot move far without destruction.

Searching by trench.—If the preliminaries are settled, and a search for a mineral deposit has been actually determined upon, there are various ways of finding its locality. If the general position of a vein, its direction, and inclination are known, it may be sought by means of an open trench, which should be drawn in such a direction as will probably cross the vein at right angles. It is also preferable to dig the trench on a hill-side, or where there is the thinnest covering of loose earth upon the rock. The trench may be made very narrow; of capacity just sufficient to admit one workman. Where it touches the vein it should be wider, in order to show to greater advantage the direction and nature of the vein.

Searching by a shaft.—Another mode of ascertaining the existence and direction of a vein is, by digging a shaft. A shaft is a vertical round pit, of from two to four feet in diameter, which is sunk down until it reaches the vein, and is generally driven entirely through it. Such a perpendicular pit is usually sunk where the mineral is covered by a heavy body of earth, or where the ground is level, and the vein lies so nearly horizontal, as to make it a profitable mode of investigation.

Searching by drift.—If the position of a vein is nearly or quite vertical, and it is so much covered by earth, as to make an open ditch out of the question, a shaft is first sunk near the vein, down to the solid rock, and into it, and from this shaft a transverse, horizontal cut—tunnel—is made towards the vein, in the shortest direction in which it may be reached. Drifts are laid horizontally or inclined, according to circumstances, care being always taken to free them from the water, which may accumulate and retard the work. In exploring a mineral region by a gallery—drift—it should be always so directed as to cut through the strata at right angles, and traverse most of the rock in the shortest possible distance.

Searching by boring.—If the direction and inclination of a vein is but faintly understood, and its locality almost unknown, the speediest mode of ascertaining its whereabouts is by sinking a bore-hole in a place where the prolonged plumb-line would be most likely to reach the vein, at the shortest possible distance from the surface. Boring is not a cheap operation; and if a greater depth than fifty or sixty feet is necessary, it will be more profitable to sink a shaft. A bore-hole cannot be made of a sufficient capacity to afford a proper examination of the vein; it merely serves to indicate, by the pounded rock—bore-meal—the general character of the rock and veins it penetrates and passes. The operation of boring for minerals is conducted in the same manner as boring an artesian well; salt wells, made in this manner, are very common throughout the Western States. Boring an artesian well, as it is performed in the West, by the aid of a rope and a heavy spindle, is an extremely simple operation, yet one requiring considerable practice; we shall not, therefore, enter into a detail of the manner of performing the operation, as mere description will not supply the place of experience to those who undertake it. In soft rock, such as the coal strata, a 2½ inch well may be bored to the depth of one foot, for one dollar, exclusive of the expense of erecting a suitable building and scaffold, supplying tools, rope, and, in many instances, a steam engine. A shaft, thirty feet deep, may also be dug for a cost of one dollar per foot, if the ground be not too hard; when the shaft runs through a rock, or is sunk to a greater depth than thirty feet, the cost per foot varies from two to four dollars. The shaft is, therefore, the cheapest in every instance, where the depth does not exceed fifty or one hundred feet.

TOOLS.

The miner has need of a variety of tools, which vary more or less in their form, in consequence of individual caprice, the kind of material they are composed of, and the nature of the mineral to be worked. Above all things, the miner needs a pick, or pick-axe. Where there is open work to be performed, a tolerably heavy common pick is made use of, pointed at one end, and having a chisel-edge at the other; in soft ground, a mud-pick, or mattock, is used. The coal-digger uses light picks, weighing 2½ or 3 pounds, to enable him to make rapid and light strokes. The miner of iron ore uses a pick of 3 or 4 pounds weight, pointed at both ends; and the miner of lead, copper, and other hard ores, uses a pick with a poll. This pick has one end made in the form of a light sledge-hammer, the other end is about 8 inches long, and pointed. This tool answers the double purpose of pick and sledge. The points of picks should be composed of the best description of steel; the economy which substitutes an inferior quality for this purpose, is misapplied.

In addition to the pick, which is the most important tool, the miner has need of a pair of iron wedges, which are, in most cases, made of steel, from 4 to 8 inches long, and hardened both at the head and edge. These wedges are driven in crevices and cracks, to break down a mass of minerals, which has been previously undermined. Each miner has a shovel, which

16

should be made of good steel, light, sharp, and pointed, so that it may easily penetrate the mass of rubbish. Shovels, with long handles, are useless in subterranean mines. A sledge, or mallet, weighing five or six pounds, is also needed in a mine; but there is no necessity for each miner having one.

Blasting tools.—When a mineral is so hard and compact, as to afford no chance of penetrating it by pick, gad, or wedge, it must be blasted; for this purpose, the miner will need blasting tools. In open works, such as stone quarries, and spacious mines, these instruments consist of drills, a needle, scraper, and, frequently, a tamping and a claying bar. A drill is an iron bar, either round or octagonal, one inch in circumference, thicker in the middle than at the ends, to prevent as much as possible the vibrations of the bar. The bar is 4½ to 5½ feet in length, sharpened and steeled at both ends, forming chisels, more or less rounded on the edge. This drill, or boring-rod, is grasped with both hands, held perpendicularly, and driven into the rock with rapid strokes. With each stroke the drill is slightly turned, so as to form a round hole. A cavity is chiselled in the rock by hand, in which the drill is started. When the drill has been driven a few inches into the rock, the hole is partially filled with water, and a ring of stiff leather fitted upon the bore-hole, with an opening in the centre of it, for the drill to pass through. This leather forms a cover, and prevents the splashing of mud and water, which would otherwise prove very disagreeable. With such a drill, holes varying from three to six feet and more in depth, are bored into the rock with the greatest ease. This drill is useless for boring horizontal holes; those having a greater or less degree of inclination may be bored with it. In narrow, low rooms, or for horizontal blasts, the long drill cannot be used in these cases; the short drill and mallet must be substituted in its place. The hand-drill, or short drill, is from six inches to four feet and upwards in length, as circumstances require; and is a bar of iron, sharpened at one end, and having a steel head at the other. This drill is held with one hand in the direction of the hole, and struck on the head with an iron mallet held in the other: the mallet varies in weight from two to four pounds, according to the weight of the drill, and the hardness of the rock. If the drill is heavy, and a deep hole is required, two men are employed to drill it; it is moistened with water, and treated precisely as in the other instances. The mallet is a short-handled hammer, having two heads or flat ends. While the boring is in progress, the hole gradually fills with bore-meal, or, which amounts to the same thing, the water is transformed by it into a stiff mud; this mud must be removed by a scraper. A scraper is a small half-inch iron rod, having at one end an eye, to serve the purpose of a handle, and at the other a flat shovel, bent rectangularly to the rod, so that it will take up the sand in the bottom of the hole, and in raising it, bring up the mud above it; forming a piston, which allows the water, but very little sand, to pass through. The miner also requires a tamping bar, in case he does not prefer to load with sand. A needle or nail, is a tapered iron rod, one-fourth of an inch in circumference, having an eye at one end: copper is preferable to iron for this purpose. It is of various lengths, from one to five feet, gradually tapered from head to point, to facilitate its extraction after the blast is loaded. Where sand is used in blasting, no needle is required. Small paper tubes, or, more commonly, long rushes or wheat straw, are used for communicating fire to the load; they are filled with fine hunting or rifle powder, and dropped into the vent-hole. The upper end of the rush-tube is split, and a piece of amadu or tinder, or a scrap of paper, rubbed over with moist saltpetre or gunpowder, is inserted, so as to communicate with the powder in the interior of the pipe. This smift, as it is called, is calculated to burn sufficiently long, to afford the workmen a chance of retiring to a place of safety.

Miners also have need of timber and an axe, to prepare props, &c. in case there should arise a necessity for their use; and wheelbarrows, and planks to run them on. The barrows are made simply of $1\frac{5}{8}$ inch pine planks, so that they will be light; and low, to save carriage, and to facilitate the loading and unloading. Canal barrows may do, but are rather heavy. For sinking shafts, a hand-whim is needed, and a kibble, into which the rubbish, water, and ore are emptied; a hemp-rope—manilla is the best,—one inch thick, will be required for raising the kibble to the surface. If the shaft is deep, and great quantities of minerals and water have to be raised, a horse-whim must be provided; and where this proves to be insufficient, a steam-engine may be employed. Crow-bars, of various sizes, are useful instruments, and should be found in every mine and quarry.

BLASTING.

Gunpowder is the most effective agent in quarrying rocks of any description, possessing an almost unlimited power. All rocks may be advantageously quarried by blasting, if they are free from fissures, cracks, and stratification; the latter does no harm, if the joints of the strata are solid, and well cemented together. A particular kind of gunpowder, called "blasting powder," is used for blasting purposes. It is coarser than common powder, and differs somewhat in its composition. It is asserted that powder may be saved by mixing it with dry saw-dust, or by ramming the sand down loosely, that is, leaving a vacant space both above and below the powder.

All the hard minerals which cannot be broken or uprooted by the pick, must be blasted. The mode of operation in blasting rocks or minerals, is very simple; still it requires caution, to avoid accidents, which, it is to be regretted, frequently happen. The first thing necessary in blasting rocks is, to select a spot proper for drilling the hole; the next, to decide on the depth and direction of the hole. The hole should never be drilled where the rock is split, as the force of the blast will be neutralized by the open fissures or crevices, and produce but little effect. A blast in slate or soft rock has but an inconsiderable effect. When the advantages of a blast are generally apparent, a spot should be selected, where it can be most advantageously applied. If the object is, to throw out a large mass of minerals, the blast should be applied in such a manner as to cause a sudden and violent shock; a heavy charge of powder, with an empty space above or below it, will accomplish this object. If it is designed merely to loosen a few large blocks, a small charge of powder will answer better than a heavy one. In every case, the amount of the charge must be proportioned to the bulk of the rock to be removed; and one of the practical lessons which miners and quarriers are necessarily obliged to study is, the method of breaking up the largest amount of rock, with the smallest possible quantity of powder. The hole is not always either perpendicular or horizontal; the direction of it depends entirely upon the species and position of the rock. If the rock is of such a nature, that it can only be removed by blasting, the holes should be drilled in such positions as to remove a certain portion of the rock, and afford a chance for the next blast; so that a succession of blasts, well applied, may separate the largest quantity of rock.

When a hole has been drilled to the proper depth,

it should be cleared of the sand and mud. The first may be withdrawn by the scraper; the latter should be removed by the swab-stick, which is a stick of dry wood, mashed at one end, so as to form a bundle of fibres, like a broom or brush. While the swab is being used, dry sand or clay—the latter is the best—should be thrown into the hole, to absorb the moisture and dry the hole. It must be perfectly dry before it is charged with any powder; and if water cannot be kept out of it, the charge must be placed in a tin tube, made perfectly water-tight, which must be inserted in the hole. The quantity of powder necessary to be used for one blast is variable, and depends on the depth and quality of the rock. Under ordinary circumstances, two inches of powder to one foot depth of hole, is used, and may be considered a fair charge. When the charge has been properly placed, the hole should be filled with small stones, sand, or clay, at the option of the miner.

Most generally, small stones are used for this purpose, which are selected from some rock that does not produce fire when struck by iron or steel; such as limestone, clay-slate, red iron ore, or, in fact, any mineral free from siliceous matter, and of a soft character. The needle or nail should be put in at one side of the hole, and pushed down into the powder; a small quantity of paper being placed on top of the powder, the hole may be filled up with the stones, which, broken into half inch pieces, are driven down by the tamping-bar and a sledge-hammer. The tamping-bar is simply a round rod of iron, three-fourths of an inch in diameter, having a swelling at the lower end, in which a groove is cut, so as to fit around the needle. By means of this bar and a hammer, the hole is firmly filled with stones. The needle, having an eye at the upper end, through which an iron rod or a short crow-bar may be pushed, is withdrawn by striking under that bar. A small hole is thus left by the needle, through which, by the aid of a rush, fire is communicated to the powder. Instead of inserting a rush tube, filled with powder, the powder may be poured directly into the hole left by the needle, to which the match may be directly applied.

When the tamping-bar and needle are made of iron, many accidents occur from premature discharges. The tamping-bar, as well as the iron needle, are apt to strike fire from any grain of sand, sufficient to kindle the powder. Needles have been, and still are made of copper; but, as the needle must necessarily be slender, it will not resist the tamping, becomes bent, and is not easily withdrawn. Tamping-bars have been made of copper, bronze, and other metals, with but poor success; they are safer than iron, but not strong enough to withstand the hammering which they receive. Iron rods, tipped or capped at the lower end with bronze, form the best tamping-bars.

Besides the improvements in tamping, in bars, and in needles, various contrivances have been resorted to, with the object of diminishing those dreadful accidents, that so frequently happen. One of the most successful improvements in blasting, rendering the operation perfectly safe, is that of using sand for the tamping, instead of fragments of rock. When sand is used, it is only necessary to fill a straw stem with fine powder, place it upon the powder in the hole, letting it project a few inches above the surface, and fill the hole up entirely with dry sand. The straw may then be cut off, close to the rock, the smift-paper or slow match fastened to it, and a small quantity of powder laid around the straw. This mode of blasting is only applicable when the holes are nearly or quite vertical; but, as most of the holes are drilled vertically, it can be generally used. It is objected, that this mode of charging requires more powder than the other; but, this is a secondary consideration, where the protection of human life is in question. Another objection which has been raised, that it is inefficient in certain kinds of rock, is totally unfounded; we have seen it used on a great variety of rocks, and have never known it to fail, when properly applied. Tough, plastic clay may be substituted for sand; but the latter is quite as efficient as the former, more easily procured, and can be filled into the hole more readily. Miners generally oppose the use of sand; the most plausible reason that can be assigned for their opposition is, that they pay for the powder, and endeavour to make the most of it. When used in short holes, sand will be found less efficient than fragments of rock, and will require a little more powder;

but the additional safety which its use guaranties, should be considered more than sufficient compensation for a small quantity of powder.

MINING.

Dead work—preparatory work.—Mining generally consists of two distinct descriptions of work—"dead work and extraction,"—the first is necessary in opening a mine, making it accessible, and advantageous for the extension of future operations; the latter, is the process of raising the ore or mineral. In addition to the above distinctions, there are "open excavations," and "subterranean excavations," each of which requires a particular knowledge, and a distinct treatment. The direction, as well as the execution of dead work, requires more intellect, knowledge, and skill, than is necessary for the raising of minerals. It may be said, that it is not the business of every miner to sink a good shaft, or drive a straight drift. The preparatory work consists in sinking shafts or pits, driving galleries or drifts, or in excavating drains for the discharge of water. All dead work is designed to conduct the miner to the most accessible part of the mineral vein, where it can be most advantageously worked; it is necessary also for opening chambers in the excavation, and in the execution of works for promoting the circulation of air, and facilitating the transportation of minerals. The inference deducible from this explanation is, that the dead works of a mine are very expensive; they are, no doubt, in many cases. When an important mine is about to be opened, true policy will dictate the expenditure of both time and money, in ascertaining the best plan of operation; the one adopted will, of course, be that which will furnish the mineral at the lowest nett cost. Before a mine is opened, a proper plan should be adopted for conducting its operations, and a calculation made of its probable consequent expense. Without a system of this kind, unexpected expenses may be incurred, which may overrun the amount realized from the sale of the minerals. The requisite length of a drift, and depth of a shaft, as well as the kind of rocks to be operated on, must be known, in order to form an estimate of the probable cost of the work to be performed. A shaft may be sunk and timbered at a cost of $5 per running yard; but it may also cost $50 per yard, owing to the nature of the material to be excavated. A level or drift—gallery—may be driven, for $3 per running yard; and, the same amount of work, under other circumstances, may cost $25. All these matters should be settled, before operations are commenced.

If a mineral vein is situated in a mountain, and it inclines towards a vertical line, with more or less deviation from it, the miner should commence by making an opening in the side of the mountain, at a place where he can effectually drain off any existing deposits of water. He should then drive a gallery, or level drift, in the direction of the vein of ore, which may serve, meanwhile, for the discharge of water from the works, the transportation of the mineral, conducting fresh air to the workmen, and as a means of exploring the deposit throughout the ramifications of the rock. From this drift, the real mining operations are extended; from this dead gallery, shafts or other galleries are extended, according to the slope, or the position of the deposit.

To excavate a gallery or drift.—The first thing necessary in locating a drift towards a mineral deposit, is, to find the lowest possible point from which the mineral can be reached. If the vein has a vertical inclination, and its extreme depth is unknown, it is the rule, to commence always at the lowest water-level, in order to drain the mine as much as possible. If the level is constructed merely for the discharge of water, it may be made narrow, but must be high enough to permit a miner to stand erect while at his work. The roof may be in the form of an arch, and the sides considerably sloped, to save the expense of timber and propping. When the gallery is intended to serve as a drain, a ventilator, and an avenue for the transportation of materials, the height and width must be increased; a width of four, and a height of five feet will be sufficient, if there is no timber required; if timbering is necessary, room must be left for the timber. A drift is always proportioned to the amount of work to be done in the mine; and the larger the dimensions of the drift, the cheaper will be the cost of transportation. In a great many mines a horse and wagon are used for conveying the minerals to the mouth of the mine, which will require the drift to be made at least seven feet high, and five feet in the clear at the bottom. If it is designed to discharge water through the gallery, a drain should be cut in the middle of the floor, and covered by cross-timbers and planks. The amount of water to be discharged will regulate the size of the drain; it should never be constructed upon a dead level, as the water always carries with it more or less sediment, which requires a strong current to float it away. One-fourth of an inch fall in the yard, may, in most cases, be considered sufficient.

Timbering a drift.—This must be done, where the rock, mineral, or ground, has not sufficient cohesion to sustain the pressure of the superincumbent mass. If only the roof of the level requires support—the side walls being strong enough to bear the weight of the roof, without bulging out,—the upright timbers should be placed against the walls on both sides of the level, and connected at the top by a cross-piece, which, while it serves to keep the uprights in their proper positions, rests firmly upon them, and forms a support for the weight of the roof. The uprights should be always slightly inclined towards each other at the top, so as to reduce the length of the cross-piece, and, consequently, increase its power of resistance. This kind of work requires a hard, solid, rocky floor, or the uprights will sink into it. If the uprights are placed against the side walls, the drift must be made proportionally wider; if they are let into the side walls, the width of the drift need not be increased. When timber is scarce, the uprights may be dispensed with, and the ends of the cross-pieces let into the rock, and rest upon it. This plan is not a very safe one; for, the ends of the timber are liable to decay sooner in the holes of the rock, than if free from it: and as the decayed cross-pieces cannot be replaced by new ones, without endangering the roof, it is rather an unprofitable mode of timbering. Sometimes, particularly when the strata is nearly, but not quite perpendicular, only the roof and one wall need support; in this case but one row of uprights is used, the other end of the cross-pieces resting in holes made in the rock. If the floor of a level is sufficiently hard and firm to support the timbers, without having the surface broken by their pressure, and the side walls and top are not quite safe, the frame should be fitted at the top, with particular reference to the side pressure. The cross-pieces should be either placed between the uprights, so as to withstand the lateral pressure, or be cut out a little, so that they may rest on and between the uprights, and have a piece of two inch plank nailed in below them. The plank should be of sufficient length to keep the uprights at the proper distance from each other. When this plan is rendered necessary, the side walls of the drift should be sloped at least a foot more than usual. If the floor is neither strong nor hard enough, to support the pressure of the timbers, a complete frame should be made, that is, with cross-ties both at the top and bottom. This frame can be so arranged as to sustain the pressure upon it from any quarter. The miner puts up the timbers or frames as his security requires them; if none are needed, he will not use any. In friable and splintery rock, he puts them in as he advances, both at the sides and top, and also sills, if needed. If the rock or ground is of such a loose character, that timber frames alone will not support it, facing-boards should be used to keep the loose stones and ground behind the timbers. Facing-boards may be simply one or two inch plank; slabs are commonly used. Split rails are the best material that can be used for this purpose. In ordinary cases, the facing-boards are put in, and the spaces between them and the rock filled up as the miner advances. In loose soil, such as gravel, loam, or sand, he puts in the facing-boards in advance, either by driving pointed planks into the ground from the nearest frame, or by cutting a small channel for the plank, and resting one end behind the frame and the other in the ground. The excavation is then made, and a frame of timber put in as soon as possible. The timbering of a drift, and all similar timber-work, is generally included in the miner's contract. The timber is delivered to him at the mouth of the mine, ready split or hewed, and cut into the proper lengths.

The size and kind of timber used in mines is a matter of consequence, as upon that and the form of the timber depend, in a great measure, the durability and strength of the work. The size of the timber should increase with the length; that is, if the dimensions of the drift require a longer timber, the thickness should be proportionally increased. A loose character of ground or rock also requires the use of thick timbers. There is no need of squaring the timber; if it is hewn at all, it may be flattened on two sides, and merely the bark and sap taken off. Rotten places in timber should be cut out, or the timber rejected altogether. Pine wood does not last long in mines; it decays in a few years. White oak may last twenty years, or longer; locust is almost indestructible. Of all woods, pitch pine is the worst; hickory and chestnut are not much better. Young timber, such as sapling, and that which grows rapidly, is scarcely worth the labour of putting in; it may serve a temporary purpose, but will not suit in works of a permanent character. Timber lasts longer in a current of fresh air, than where the air is foul or damp.

Drainage.—The drain for carrying off the water from a mine, should be the first thing commenced after the drift has been started. A drain is laid beneath the floor of the drift, at such a distance from the surface as to leave room for a cover, and, if possible, for a good layer of rubbish from the mine. If there is any fall from the drift to a valley below it, a sloping, descending direction may be given to the drain. In soft grounds drains are most frequently made of planks, spiked together in the form of a square trough, which is buried below the surface. This is bad policy, as these troughs are liable to be crushed by the weight of ground generally thrown upon them; in this manner, the drains become choked up, throwing the water either back into the mine, or over the platform before it. Both evils cause a disturbance of the regular operations, which should be avoided. A drain constructed in soft ground, particularly where it is deeply buried, and not accessible, should be built of stones, which may be laid without any mortar. There should be first a pavement of stones laid, then side walls of the same material may be constructed, and the whole covered by stone slabs. Such a drain is very durable, and not easily crushed by the superincumbent weight.

Our mining operations being yet in their infancy, there is but little inducement offered for the erection of permanent works. Temporary improvements may be in many instances the most proper; still, there are cases where solid, durable improvements are profitable. Where a drift opens an important and extensive mineral deposit, with a prospect of its continuing profitable for many years, economy would dictate to use stones instead of wood, in constructing the underground works, and adding to their safety and permanence, particularly where good and durable timber cannot be obtained. In brittle rock and loose ground, if the drift is a large one, timbering is not very cheap. In these cases, it requires strong and close timbering; the timbers should be eight inches thick, and the frames placed close together. The usual distance between the frames is three feet; which in soft ground is reduced to two feet, and, in slaty rock, increased to four feet. A soft floor, with a heavy roof, composed of fragments, requires a larger quantity of timbers, and of greater strength than usual. This frequently occurs in coal and iron mines.

The excavation and timbering of a drift, is a kind of work, which affords the miner an opportunity of displaying his skill, as well as his dexterity. In driving a level, a remarkable difference will be apparent in the abilities of different miners. While one description of workmen advance but slowly, others will make a rapid progress. As the width of a drift will not allow of the employment of more than one or two miners at the same time, it becomes a matter of importance to place the work in the hands of the best operatives. An uneven floor, irregular timbering, and a crooked drift, are evidences of the unskilful workman.

Cross-ties or sills — sometimes called sleepers — are laid in the floor of the drift, upon which iron rails or pieces of scantling are fastened, forming a sort of rail-road or tram-way, on which the minerals and rubbish may be transported to the mouth of the mine. Where no heavy wagons are used, and but little material transported, white oak lath may be substituted for iron rails. Sometimes a dog-cart, with two high wheels, and a boy to act as driver, is employed; in which case a plank floor is laid, for the whole to run upon.

Shafts.—If drifts are considered to be horizontal excavations, shafts may be said to be vertical ones. There is, however, more variety of direction given to shafts than to drifts; shafts are usually made vertical, but they are also given a greater or less degree of inclination, according to circumstances. If the position of a mineral vein will admit of a shaft being laid in its plane, it may be found profitable to do so. If no other advantage will arise from an inclined shaft, than the mere value of the mineral raised during the sinking of the shaft, that will be but a small inducement to give it that direction; for, what may be gained on the ore, will be overbalanced by the greater quantity of timber and other materials used, the additional length of water-pipes required for the pumps, and other inconveniences, resulting from the extra length given to the shaft by its inclination. Cases may occur, where it would be advisable to sink an inclined shaft, but they will be very rare. An inclined shaft may be found profitable, if it can be sunk with sufficient slope to make the use of timber unnecessary. When steam or water-power can be applied at a certain point, the transportation of which to another point would cost too much, an inclined shaft may be found serviceable. It may also be used where the mineral cannot be reached in a shorter way. There may be advantages in a sloping shaft, under peculiar circumstances; but they should be well considered before the plan is finally adopted. The primary expenditures are one item in such calculations, but the cost of transportation is another, which generally overules the first.

The sectional dimensions of a shaft depend chiefly on the amount of minerals to be raised in it. If the quantity of minerals is not so great as to require a double shaft, a single one may be sunk; 4×4, or 4×5 feet, in the clear, is sufficiently large. Double shafts are made, 4 to 5 feet one way, by 8 to 12 feet the other way. A partition throughout the entire length of a double shaft divides it into two shafts of equal size. If pumps are required in the shaft, a division is made for them, either at one of the small sides, or in the centre of the shaft. If a shaft is merely intended for the ventilation of the mine, any form or size will do; the chief object being, to secure it against the danger of falling in, in consequence of the timbers giving way. Shafts are usually made subservient to every purpose connected with mining, such as hoisting, pumping, ventilation, and the ascent and descent of the workmen. When used for these various purposes, the shaft should be strongly timbered, and secured against accidents from outward pressure.

The timbering of a shaft requires more skill than that of a drift. Shafts are made of a rectangular form, because it is more convenient for the miner, and renders the execution of the wood-work more safe and easy. The timber is generally from six to eight inches square, and framed in a rectangular form, the inside of the frame being the dimensions of the shaft. These frames are seldom placed close together; a space of 1, or 1½ yards, is generally left between them; but, when the lateral pressure is very considerable, and the timber cheap, but of inferior quality, such as pine timber, the frames may be placed in contact. The timbers composing a frame are united at the corners by a half-check; the longer pieces extend a foot or more beyond the frame, and rest in holes cut in the rock. When the frames do not touch each other, facing-boards, consisting of strong planks, or split rails, are fastened behind the frames, to sustain the pressure of the ground. If raising water from the mine is an object, and the surrounding ground is wet, great attention should be given to the planking of the shaft; and, if possible, the space between the planks and the ground should be filled with puddle, in order to keep as much water out of the mine as possible. The planks are fastened to the frames by spikes, or in some other way, so that the whole frame-work of a shaft forms one solid, united body of timber. Whether a shaft is perpendicular or inclined, the frame-work should always be placed vertically upon the basis or axis of the shaft, so that the frames will be at right angles to the four sides of the shaft. Inclined shafts are timbered in the same manner as drifts, in case they are not too steep to admit of it.

Shafts and galleries form the bulk of dead work;

the other descriptions of it are not of so much importance, nor do they form such an expensive item. These works are not only excavated in dead rock, but also in the mineral veins; still, they cannot be considered as anything but dead work, for the minerals raised seldom or never pay the whole cost of the work. Preparatory works are made with the object of advantageously reaching and working the deposit; also, with the view of laying open the deposit to such an extent, as will guaranty the extraction of a certain amount of minerals in a certain time. This last object is seldom properly understood or estimated by those who engage in mining. If a drift or a shaft has a sufficient capacity for the transportation or raising of 100 tons of mineral per day, it does not follow, that the mine can or should furnish 100 tons of mineral in that time. If one miner can dig two tons of mineral per day in a proper room, two, operating to the same advantage, will, in a similar room, furnish four tons per day, of twelve hours. If the work progresses day and night, one room will furnish eight tons per day, of twenty-four hours. To supply 100 tons per day, of twenty-four hours, would require thirteen rooms. A room for two miners should be at least from twenty to thirty feet wide. If the vein lies nearly or quite horizontal, and we take the most simple case — that is, locate all the rooms on both sides of the main drift — it will require a drift nearly 300 feet long to locate the thirteen rooms. If the work-rooms, or stalls, are located only on one side of the drift, which is often the case, owing to the inclination of the vein, the drift must be twice the length before mentioned. If the position of the vein is such, that it cannot be worked from the main drift, the case will be rendered still worse; branch drifts must be excavated, in order to reach the minerals. The case becomes still more complex, when the veins are thin, and almost vertical. The work in galleries and shafts cannot be hurried, without disproportionately increasing the expenses, because only a limited number of men can work at the same time. Time is, therefore, required, to open a mine properly, and put it in a condition to furnish a certain amount of minerals. It is bad policy to force the productiveness of a mine beyond its capacity; the consequence of such a proceeding will be a waste of money, or a higher cost of the product.

EXTRACTION OF MINERALS.

A full description of the numerous varieties of preparatory work is beyond the limits of this volume; we shall mention them, as we progress with our detail of the extraction of minerals. All mining operations may be separated into two distinct divisions; those carried on in the open air, or open excavations, and those conducted beneath the ground, denominated subterranean workings.

Workings in the open air present but few difficulties. Any workman who can handle a pick or a shovel, and push a wheelbarrow, is competent to do the work of a miner in open ground. If the minerals are not covered by very heavy masses of rock or earth, the work will occasion but little expense, and the minerals may be procured at a trifling cost. Open workings are generally resorted to, where the minerals are close to the surface, and where the substance to be raised is covered with matter having no cohesion, and which is, consequently, not strong enough to admit of subterranean operations. Open excavations are not profitable in general: they are resorted to when every other method proves impracticable. The masses of ground to be removed, the disturbances occasioned by the changes of weather, the accumulation of water, and the dangers to which life and limb are subjected, are among the many disadvantages of open workings. The rules to be observed in open excavations are — to lay out the terraces in such a manner as to facilitate the removal of the earth; to dig and timber drains, for the removal of water; and to lay out one or more barrow-tracks, from the diggings to the scaffold, where the ore is loaded. The transportation of rubbish must be accomplished at the smallest possible cost; the division of the bank into terraces will have a favorable effect, as it affords the means of removing the ground almost horizontally, either by the use of wheelbarrows or shovels. Above all things, the crowding of loose ground into a small space should be avoided, as it necessitates the lifting of the earth. If the ground can be removed down-hill, it will always be desirable to do so. The cost of removing a cubic-yard of earth from an open mine, or ore bank, should not be more than fifteen or twenty cents. If room is made for low horse-carts, the ground may be removed at a cost of ten cents a yard; and, if the ground is so loose and free of stones as to admit of ploughing, and scooping by a horse-scoop, a cubic yard may be removed for six cents. The overlying ground in open workings is frequently broken up by undermining the bank, above the mineral deposit, and then overthrowing it by the aid of crow-bars or heavy wooden wedges, driven in from the top. The mass of rubbish may, in this manner, be all thrown down into the lowest part of the mine; and, frequently, heaped up higher than the original bank. If the ground is of such a character as to defy ploughing, scooping, and working in terraces — as may be the case where very tough clay, gravel, or slaty rock is the matter to be removed — the bank should, when thrown down, be loaded upon low horse-carts, or, which is preferable, upon sleds. The use of high carts is as unprofitable as would be the shovelling of the loose rubbish into a heap; the lifting of ground by manual labour is the occasion of great expense. The up-hill work is, in every case, far more expensive than down-hill work. The transportation of ground down-hill, and by animal labour, should be effected, by all means, in every case where the nature of the ground will permit.

Many accidents occur in open workings from the crumbling down, or caving in of the sides of the bank, which might be prevented by giving to them a suitable slope; but, as this is not always practicable, the plan of working in terraces may be resorted to with perfect safety. If neither of these expedients is practicable, the banks should be propped by timber wherever they appear to be dangerous. The plan of stripping a mineral deposit, is seldom the cheapest mode of operation; if the work can possibly be performed by subterranean excavations, it will be found most advantageous to do so, especially where there is a prospect of a large amount of dead work, which cannot be profitably done. Under-ground works are not clogged with masses of rubbish like those in the open air; but, as somewhat of a counter-balance to this advantage, there is some dead work to be done in shafts and drifts; this, however, if assessed upon the aggregate product of the mine, will not amount to any considerable sum.

Digging of clay.—Open workings are resorted to for the excavation of loam or clay. If the latter is covered by heavy masses of impure material, rendering the stripping of it unprofitable, round shafts may be sunk; these can be rendered secure from accident by the insertion of a casing, composed of strong hoops. As much clay is taken out as can be possibly removed from one pit, and the casing of hoops withdrawn: this done, the pit is left to take its chance; it gradually caves in, and fills up the vacant space.

After the lapse of a short time, a new pit may be sunk by the side of the old one.

Valuable sand is also excavated in open workings. The stripping of a sand-bank frequently costs more than the value of the sand obtained. If the sand is situated at such a depth beneath the surface, as to make the stripping of it unprofitable—under-ground excavations being, in the majority of cases, entirely out of the question — it may be raised from a depth of thirty or forty feet, by the aid of a shovel formed for this purpose. This is an instrument, having a long wooden handle, the length of which may be increased at pleasure; at the lower end a shovel is fastened, at nearly but not quite a square angle with the handle. This shovel, or scraper, being made to revolve horizontally upon its axis, will form a round hole, like a small shaft. A shovel, fourteen inches square, will make a shaft eighteen or twenty inches in diameter. By throwing out, on each revolution of the shovel, or scraper, all the sand or earth that has accumulated upon it, a small shaft may be sunk; the rubbish may be removed, and the good sand saved, when raised by the shovel. The shaft or pit may be increased in diameter at the bottom, so that a large quantity of sand may be excavated from one opening.

Marl is usually dug in open workings, which are often of considerable depth. There are marl-beds in New Jersey, over fifty feet below the surface, which, as may be imagined, cause a great expense for the stripping. In these cases, expensive subterranean operations will be found more profitable than open workings; there being usually an overlying rock, sufficiently strong for a roof.

Most of the iron ores are dug in open workings. There are, in this country, such heavy and extensive deposits of iron ore near the surface of the ground, and buried in alluvial clay, that extensive subterranean operations are not profitable. Generally speaking, one foot of stripping may be done to obtain one inch of ore; that is, a vein of good iron ore, twelve inches thick, will repay the cost of stripping it of a mass of matter twelve feet thick. This rule, however, is so vague, that but little dependence can be placed in it. The kind of ground to be removed, and the facilities furnished for that purpose by the locality, will, of course, regulate the amount of stripping that may be advantageously performed.

Limestone, building-stone, gypsum, roofing-slate, alum-slate, and some other minerals, are mined or quarried in open workings. These minerals are generally found in such heavy masses, that the cost of stripping a quarry is trifling, compared with the quantity of minerals taken from it. Gunpowder is extensively used in quarrying, when the size or shape of the stones is not a desideratum; this is the case with limestone, gypsum, and the ordinary building-stone, used for rough walls. When blocks of stone, of a definite form and size, are wanted, the use of gunpowder must be dispensed with, at least, after the opening of the quarry. Building-stone is quarried by drilling holes in a line, at the place where it is intended to separate the stone from the mountain rock. These holes should be deep or shallow, according to the value of the stone, and the facility with which it breaks in the direction marked. Large blocks of marble are usually quarried by boring the holes close together, and through the entire thickness of the stone. If the quarry is in the form of a terrace, but two sides of the block will require drilling. In stones of less value, the holes may be made from three to six inches deep, and at a distance of from six to twelve inches. In these holes pieces of dry, hard wood, are driven, which usually crack the stone in the whole marked length, at the same moment. Iron wedges are commonly used for this purpose; they are not driven directly into the holes; two iron cheeks are inserted in each hole, and the wedge driven between them. Iron plugs or wedges do not work regularly; it requires more care to drive them uniformly, than is the case with wooden plugs.

Gold, diamonds, platinum, and a variety of other minerals, are mined in open workings. In California, Virginia, and other States of the Union, when gold is found in connection with gravel, sand, or soil, it is deposited in rockers, in which the mud, sand, and gravel is washed away, and the coarse gold deposited in the bottom of the machine, where it is either found in grains, or amalgamated with quicksilver.

Anthracite coal is, to a great extent, mined in open workings; as are, also, turf and lignite.

Subterraneous workings are all conducted on the same principle. Some slight modifications may be made in practice, but they do not interfere with the general arrangement. We will, for the sake of imparting clearness and perspicuity to our descriptions, divide these operations into various classes.

Horizontal veins, or such as are nearly horizontal, and not more than six feet in thickness, are generally the most convenient, and can be worked at the least cost. If such a strata is situated in the side of a hill, above a natural drain of water, it may be explored by a drift, which should be directed towards, or commenced at the lowest part of the vein. If it is mineral coal, and the drift be located in the vein itself, it should be commenced in the warm season, and prosecuted to completion in the winter, on account of the greater rapidity with which foul air is generated in summer than in winter. While the drift is in progress, a shaft may be sunk above the vein, and through it, which will either meet the main drift, or a branch extending from it. This shaft forms a chimney, which carries off all the foul air from the mine. The deeper the shaft, the stronger will be the draft created by it; so that a lively circulation of air may be produced in this manner. An air-shaft should be so located as to ventilate all the work-rooms, throughout the entire extent of the mine.

The working of horizontal veins is accomplished by various methods. If the vein is thin, and will not allow of much dead work; or, if the means at disposal for the performance of such dead work are limited, the drift should be made narrow and low, and so driven as to be in contact with the vein throughout its entire length; the vein should be either at the roof or the bottom of the drift. The drift should also be so constructed as to discharge all the water which may accumulate in the excavation. This level or drift may be carried as far as convenient and safe, say, 50, or 100 or 200 yards; bad air, change of ore, or want of means, being sufficient reasons for not pushing it any farther. At the extreme end of the drift, the miner commences to extract the ore, taking out as much mineral from each side of the drift, as he can do with safety and profit. If the roof is strong, that is, composed of a solid mass of rock, without fissures or scales, he may excavate to the distance of ten or more yards on each side of the drift, opening a room, twenty yards in width. The miner works back towards the mouth of the drift, discharging the minerals by means of wheel-barrows or rail-road cars, as the drift is too narrow and low to permit the use of a horse. If the vein is thin, he will be obliged to excavate a sufficient amount of dead rock, to give himself room to work. In these cases, the work is performed by the miner, while lying on his side; and the amount of room which he makes for himself depends, in a great measure, on the opportunities he possesses of undermining the vein. A seam of soft material is generally found near a vein of minerals, either below it or above it; it works to better advantage if below, and is called the undermining. The want of this undermining is often the reason for abandoning a small seam of minerals; for, its extraction in hard rock would cost too much. If this undermining is near the vein, the miner will not, of course, take out any more than the mineral and undermining, if it will afford him sufficient room for working. If the undermining is more or less distant from the vein, he is obliged to take out all the material between it and the vein. The rubbish, or fragments of rock made in the progress of the extraction, are thrown behind, to fill up the space between the floor and the roof. This facilitates the work, and renders the miner more secure; for, if the roof has not sufficient strength, it may sink, but cannot break down upon him. If the vein is thick and pure, and furnishes no material for the support of the roof, and the roof is weak, wooden props should be used for its support. The roof is generally allowed to break down behind the miner as he advances; but it requires experienced workmen to judge by the appearance of the rock, when the roof is about to cave in, and be ready to provide for their safety accordingly. It requires a good ear to detect the sound of the moving rock, and to be able to judge of the time when it will fall in. A peculiar, crackling noise, always precedes a fall of rock, which is never allowed to pass unheeded by the experienced miner; this warning sound gives him time to escape from danger. All miners who are not sufficiently acquainted with these indications of peril, had better prop the roof. This would be advisable in every case; or, if props are not sufficiently secure, pillars of ore should be left standing in places where the roof is weak, where fissures exist in the rock, or where loose blocks have a threatening appearance. In this way, nearly the whole amount of minerals may be taken from a mine, so far as the extent of the drift will allow. If, after the removal of that part of the vein, it is desirable to penetrate farther into the hill, another and more permanent drift should be made.

If more permanent works are designed, and it is desirable to exhaust all the minerals contained in a certain field, solid and commodious preparatory works should form the basis of the operations. A drift, in this case, should have a height sufficient for the admission of a small horse, to be employed in drawing out the loaded cars, and returning with the empty ones; it should also have air-shafts and water-drains, properly constructed. A system of works should be laid out, sufficient to secure the future drainage and ventilation of the mine, and the profitable extraction of the ore therefrom. From the main drift, branch drifts may be extended for the exploration of the mine, and the increase of its capacity; or work-rooms or stalls may be excavated from the main drift in such directions as will secure dry works. The first object to be attained in a mine is, to have dry rooms. If the water cannot be drained from the work-chambers directly, drains or drifts must be excavated which will secure permanently dry rooms. These chambers are made from ten to twenty yards in width, according to the strength of the roof. The more room each miner can procure, the more profitable will be his labour; therefore, a width of six or eight yards should be given to each room intended for the use of one miner. Chambers may be driven from the main, or any other drift; but, if the roof be weak, a mass of mineral, equal to one-third of the size of the chamber, should be left standing between adjoining chambers, as a support for the roof. These supports may be partially removed after the mine has been exhausted, by commencing at the extreme end of the mine, and cutting away everything but a few square pillars, which must necessarily be left, to insure the safety of the miner; the weight of the superincumbent strata will, in a short time, however, crush these frail supports. Bold and dexterous workmen often succeed in removing all the minerals, leaving the roof to fall as they retreat.

Veins more than six feet thick cannot be excavated at once. The drift should be located in the lowest part of the vein; and, if the roof be sufficiently strong, chambers of various capacity may be opened throughout the whole extent of the mine. If there is but a weak roof, the chambers should be small, and the extraction of the mineral should commence at the extreme end of the vein; in this case all the mineral must be extracted at once.

If a horizontal vein is situated beneath a plain,

rendering a natural discharge of water impossible, it will be necessary to sink two shafts, at a certain distance apart, which should, if possible, touch the vein at the same degree of inclination. A drift may be driven through the vein to unite the two shafts. One of these may serve for an air-shaft; the other, for draining and hoisting purposes. The latter shaft should be placed at the lowest part of the vein; the former, at the highest point. From one of these shafts, which forms the mouth of the gallery, the work may be prosecuted, as in other cases; the only difference being, that machinery, with either steam or water as the motive power, will be required for the removal of the water and minerals from the subterranean workings.

An ore or mineral deposit of great thickness should not be pierced by shafts. Both the shafts and road-levels should be at a distance from the mineral, so that none of it may be lost by being left standing in the mine as a support for the shaft and machinery, which would, otherwise, be exposed to the danger of falling into the mine. Heavy masses of mineral are always pierced by galleries, and worked from behind, so that the minerals may be all extracted before the roof caves in.

Veins having nearly a vertical inclination, should be worked in a different manner from horizontal veins. Arrangements must always be made for ventilation and drainage, no matter how the vein is attacked; whether it be by a level driven from the lowest and nearest point, or, by shafts, sunk upon the vein, either perpendicularly, or in the line of its plane. A drift may be driven through the vein from the dead or drainage gallery, to a shaft some distance from it, or from one shaft to the other. If the vein is thick enough to admit a workman, nothing but pure ore should be taken out; but if it has not the requisite thickness, enough dead matter should be extracted with the ore, to allow the miner room to operate. Similar galleries may be driven from the level, or from the main shaft, in both directions, through all the ramifications of the vein. The lower galleries serve principally for the purposes of ventilation, drainage, and the transportation of the ore. At a distance of twenty yards or more above the first galleries, another series may be opened, horizontal to and parallel with the first, and connected with the same air and main shafts. In this manner galleries are opened, until sufficient room is obtained for the accommodation of the number of men required for the performance of a certain amount of work. That part of the vein which is situated between the two series of galleries, is divided into terraces — one terrace being six feet in height, to allow a miner to stand upright.

The mass of minerals between two galleries may be extracted, commencing either at the entrance, or at the farthest part of the gallery. It may be worked either from above or below. This may be accomplished by one miner, or, if the vein is more than four feet thick, two miners and one helper will be required. They proceed to excavate a step, six feet in thickness, either beneath the floor of the upper gallery, or above the roof of the lower gallery. In the latter case, the miners must put in timbers as they advance; in the first case, a timber floor is not required, until a second set of hands commence operations. When the first set of miners have advanced some six or eight yards, a second set is put at work, either above or below the first. When both sets of operatives have excavated a further distance of six or eight yards, a third terrace is commenced, and so on, until the whole space between the two galleries is occupied. If the work has been commenced from above, the whole will have the appearance of a pair of stairs with very large steps; if the work has been commenced from below, the steps will have an inverted appearance. Each step or terrace is prolonged by a strong floor, composed of stout timbers let into the rock, and planked; this floor serves for the transportation of the ore from each terrace to the hoisting-shaft, or the car-level. It also affords a space for heaping up the rubbish taken from each terrace. The timber and planks composing the floor may be removed after the mineral is exhausted. The roof of the lower gallery should be strong enough to bear the pressure of all the rubbish from the upper workings. From these rooms or terraces the ore vein may be followed in all its ramifications through the rocks, and every particle of mineral be removed, before the place is finally abandoned.

When working with inverted terraces, it may happen, particularly in thin veins, that a great deal of otherwise troublesome rubbish may be used to advantage. If the roof of the lower drift be strong, and able to bear a good load, the rubbish may be left on the spot where it accumulates, forming a slope, which may be used as a substitute for a scaffold. In this way a great deal of timber may be saved, as this plan of working requires very little, if any. The ore is shovelled down the slope, and passes through the roof of the lower gallery, from which it is removed to the cars. This mode of operation answers very well for a coal mine, but not for any valuable ore; as a great deal of mineral is unavoidably left with the rubbish, and much dead rock carried along with the ore.

Both systems of working a horizontal vein have their advantages and disadvantages. The descending terraces use a great deal of timber, but afford an opportunity of doing clean work; there is more shovelling necessary in this case, but the miner has the advantage of using the pick and drill in the best position, and to the greatest advantage: this is the method of working valuable ores. The ascending, or inverted terraces, particularly where the rubbish is used as a scaffold, is an imperfect method of working, as the ore cannot be separated from the rubbish in the mine. The natural descent of the minerals is, however, an advantage; the miner is not compelled to use the shovel for the removal of the broken ore.

If there is any doubt of the solidity of the partition walls in the vein, they should be secured by props. The same practical mode of working, used in the horizontal vein, may be here followed; that is, an undermining, or soft part of the vein, may be found, by means of which the vein may be separated from the rock, when it can be broken into fragments by the use of wedges, the sledge-hammer, or gunpowder.

Heavy masses of minerals are mined with more difficulty than is at first sight apparent. The process is as follows: A heavy body of ore is pierced by wide parallel galleries, leaving pillars or walls standing between them in the form of long arches. In this manner all the mineral in one level or story, of from six to seven yards in height, is exhausted, before the next level is commenced below it. There is a loss of mineral connected with this system of operations; but it is the only rational mode of working, every other plan involving a larger expenditure of money, and a greater waste of mineral. Any other plan may either entirely ruin the mine, or make it work to a great disadvantage. The mass of minerals left in a mine by the system of working it in stories, is equal to, if not more than one-half of the whole amount of mineral; but this is not lost, it may all be taken out at some future time, when the mine is so far exhausted, as to afford no reason for its farther extension. The pillars and arches left must be strong enough to bear the weight of the superincumbent rock.

Another method of working heavy masses, on the principle just mentioned, is, to work them from below. For this purpose, a drift is driven below the deposit; or, if it is too deep for that, through the vein itself. From this drift, the mass is divided by pillars into wide rooms or galleries. These are the working rooms, or stalls; between each two a wall is left standing for the support of the roof. All the rubbish made in extracting the ore, remains in the mine, and, as the roof is excavated, the floor is proportionally raised. In this manner the vein may be worked to an immense height; and, if sufficient dead matter is made, the pillars may be removed, as the rubbish is strong enough to support the roof. If the works be not commenced below the mineral deposit, and there still remains some ore below the lowest point to which the mine has been excavated, a floor of strong timber may be laid, for the rubbish from the upper part of the vein to rest on; this serves for a roof to the lower part, in case it becomes expedient to extract the ore from a greater depth. If workings of this kind are carried on in stories, of which the second is commenced after the first has been entirely exhausted, the whole amount of mineral in each story may be removed at once, because the rubbish forms pillars of sufficient strength to support the roof. If sufficient rubbish is not accumulated from the working of the vein, a subterranean quarry may be opened in a convenient place, from which stones may be quarried at a small expense, and transported to the mine. This is the most advantageous mode of working, not only in heavy masses, but also in either vertical or horizontal veins. It affords great facilities for making the rubbish subservient to the purposes of practical economy; adding to the solidity of the mine, saving timber, and lessening the expense of extracting the ore. If rubbish is needed for filling up, it may be procured from interior vaults, or be sent down from above, in the returning kibbles or cars. There are many other important advantages, recommending this mode of mining.

REFLECTIONS.

The miner is exposed to numerous perils while engaged in his search for mineral treasures. The rocks and other substances in which he delves are seldom entire in their formation; he almost always finds them divided by clefts or fissures, and the impending fragments threatening to fall and crush him. Decayed and friable rock, alluvial sand, gravel or loam, frequently surprise the unwary miner, rushing down upon him like an avalanche, and, in many instances, entombing more than one individual. Foul, inflammable air is ready at any moment to ignite, and either roast the miner alive, or, by the irresistible force of its explosion, tear from their position immense masses of rock, annihilating the miserable miner and his colabourers before the expiration of their natural term. The miner endeavours to protect himself from all these dangers, by using the precautions which science and experience have taught him to believe are most efficacious. Yet, all his knowledge, accompanied, as it usually is, by an acute sensibility to the intimations of impending peril, is frequently an insufficient security from harm; the elements with which he has to contend appear in such new and various shapes, as to baffle all his previous experience. The miner is subject to a greater number of accidents than any other class of operatives; he has a greater cause to look for the protection of a power superior to that of man. For this reason, we generally find the educated and scientific miner more truly pious, and the ignorant more superstitious, than other classes of men. Their avocation subjects them to the possibility of being summoned, in an unguarded moment, to render their final account.

VENTILATION OF MINES.

A free circulation of air is of the utmost importance in every mine, not only for the security of life and health, and the protection of property, but as a means of decreasing the cost of production. The safety-lamp, invented by Sir Humphry Davy, has been highly lauded for its protective qualities; but it has been found inconvenient and insufficient — accidents of the most disastrous character having occurred where this lamp was in use. Too great a dependence being placed upon the assertion, that inflammable gas cannot be ignited by the flame of this lamp, the miner becomes careless of the circulation of air, and other necessary precautions; the gas accumulates until it acquires a fearful density, when an accidental spark of fire may be the means of its ignition, and, consequently of a more horrible destruction than would have been the result, had proper regard been paid to the means of ventilation. When men are at work beneath the surface of the earth, confined in close and narrow passages, their respiration, joined to the combustion of candles and gunpowder, and the decomposition of wood and minerals, soon vitiates the atmosphere. Sometimes, sulphurous, arsenical, mercurial, and other vapours, are disengaged, and prey upon the health of the workmen; but, the most dangerous in its character, and productive of the most disastrous consequences, is carburetted hydrogen — the fire-damp of the coal mines. The only successful remedy against all these evils is, a free and liberal circulation of air. The means employed to produce this effect, have been almost reduced to a science; our purpose will not allow of a lengthy exposition; we can merely trace the outlines of the subject. There are two methods of ventilating a mine — the natural and artificial. Currents of air, produced by the difference of density between the air of the mine and that of the atmosphere, constitute the natural system of ventilation. Artificial currents of air, produced by machinery, in the form of exhausters, fans, and blast-cylinders, and by the aid of fires, and other forces, are the substitutes for natural ventilation. Artificial ventilation always entails considerable expense, and should, if possible, be avoided.

The temperature of subterranean workings is always higher than the mean temperature of the place in which the mine is located. Hence, the air of a mine in winter is more rare, and in summer more dense than that of the atmosphere. For this reason, when the mine has two openings, at different heights, the air will be found to flow out of the lowest one in summer, and the highest one in winter. This principle, properly applied, affords extensive means for the ventilation of mines. Moisture engenders vapours, which are not as heavy as atmospheric air; the waters of a mine may therefore be advantageously used for ventilation. Where a succession of shafts or drifts communicate with the atmosphere, no difficulty will be experienced in producing a current; in such cases, a natural draught will usually be produced. If this does not happen, an artificial draught may be caused, by erecting a chimney upon one of the openings, which, if of proper dimensions, will furnish a sufficient circulation. During the heat of summer, and the cold of winter, there is usually no difficulty in ventilating a mine; but, in the seasons of spring and autumn, natural means are often found insufficient, owing to the interior and exterior air being of the same degree of density, and, consequently, too feeble to produce a current. A fire kindled under a chimney, constructed for the purpose, never fails to produce a current. The worst time for ventilation is, during the prevalence of storms; stormy weather deranges every natural system of ventilation. Machinery, not being subject to natural changes, can be depended upon in any emergency; but the ventilation of a mine by such means is very expensive. Fans, or centrifugal blast machines, are the best for this purpose. The amount of fresh air necessary to the purification of any particular mine, cannot be determined by any general law; locality, material, and mode of working, have each a bearing upon such a calculation. A coal mine has been known to explode, when 5000 cubic feet of air per minute had been forced into it.

PRACTICAL REMARKS.

There is a vast quantity of useful minerals in the United States, which are so profusely distributed above the water-levels of the valleys, that there is very little prospect of any extensive operations below that level, for a long period of years; at least, so far as the general production of minerals is concerned. But few establishments conduct their operations at such a depth below the usual water-level, as to require the use of machinery for drainage purposes. Comparatively few coal mines have any need of pumps; and but few gold, lead, iron, and other mines, are troubled with water. The majority of the mines are drained by natural means. This circumstance renders mining comparatively easy and cheap, and simplifies it so much, as to enable men, who are not possessed of a practical or scientific education, to work in and conduct mines. A workman, possessed of good common sense, has sufficient ability to conduct mining operations of this kind. Our mines may, in many instances, be opened and conducted with but a trifling capital; all that is required to make a mine prosperous is, to obtain a good market for the sale of the minerals.

The mining of minerals is quite a different business from the smelting of them; and, if metallurgical operations are to be profitably conducted, they should be entirely separated from those of mining; each requiring a particular attention, and a special education and knowledge. Smelting operations can be most successfully carried on where a variety of ore is obtainable; not only of different chemical compositions, but also from different mines. Another reason for the separation of both branches is, that smelting works should, in order to make them profitable, be conducted on an extensive scale. Mines are not always in a condition to furnish a regular supply of ore, much less to produce it of a uniform quality. Smelting works should not suffer from the inconveniences resulting from a deficiency in the product of the mines. It is the interest of those who possess minerals to facilitate the erection and prosperity of smelting works, as independent establishments, by furnishing them with minerals at reasonable prices. The farmers and owners of lands are generally in a position to conduct mining on a cheaper scale than those who carry on smelting operations; and, if they furnish minerals at the prices for which they can be raised by the proprietors of the smelting works, no doubt can be entertained of their obtaining a preference in furnishing the ore. The advantages arising from the peculiar situation of the farmer, guaranties a profit which would otherwise be lost to himself, and consequently to the community.

The digging of minerals is sometimes connected with considerable expense and uncertainty, particularly where a great deal of dead work is to be done. Preparatory work is naturally expensive, and if not accompanied with certain positive results, which may be previously estimated, it will be safer not to engage in extensive experiments. On the opening of a new mine, it is, in every instance, the most prudent plan, to commence operations at the most accessible point, and open the mine at the least possible cost. If the first profits are not large, there will yet be no heavy losses; and, as time has the effect of diminishing the expense of mining operations, explorations of the vein may be successfully prosecuted while it is being worked. The extraction of minerals in this way is often found to be too expensive; but, if this be the case, it will be absurd to incur a heavier expense in the construction of shafts and drifts, where the nature of the vein is unknown.

If the nature, direction, and inclination of a vein of mineral be known, shafts may be sunk upon it with the greatest certainty, and a safe estimate can be formed of the cost of sinking them. A cubic yard of gravel, sand, or rotten rock, may be removed from a shaft, at a cost of from $3 to $6; the price advancing with the increasing depth of the shaft. In hard rock, such as that of the coal formation, a cubic yard may be raised at a cost of from $4 to $8. In transition rock, the price will vary from $5 to $10, timbering included — the timber to be delivered at the mouth of the mine. The excavation of a level drift can be performed for one-half of the above rates;

which prices are considered sufficient for the excavation of a fine, durable drift. Shafts and drifts, constructed for the purpose of examination, are generally of a small size, and made in a temporary manner. A shaft, thirty-two inches in diameter, may be sunk for a cost of $1, or $1 50 per foot, in depth; and a drift, three feet wide and five feet high, may be driven for a price varying from $2 50 to $6 per running yard. Excavations of large dimensions can be made at a comparatively less cost than those of a small size. In open workings, the stripping may be done, and the ground removed, at from ten to thirty cents per cubic yard. In all bargains with miners, the timbering of the excavations, if not particularly stipulated, is, nevertheless, understood to be included. The timber is required, in every case, to be delivered at the mouth of the mine. It is also understood, that the repairing and sharpening of tools must be done by the miners, if they are not exempted by a stipulation to the contrary. The tools are furnished to the hands, and they are charged with their price: when they return them, either broken or sound, the tools so returned are placed to the miners' credit, and they relieved from any further responsibility. Gunpowder, and other necessaries for blasting, are also furnished at the expense of the miner.

The price paid for the excavation of minerals varies according to circumstances. Iron ore is extracted in many localities at a cost of from 50 to 75 cents per ton. There are also ores mined at a cost of $4, and upwards, per ton. On an average, we may say, iron ores should not cost over $5 for each ton of iron. There are exceptions to this rule; but, the average cost will seldom be so great. In the present condition of the iron business, operators cannot afford to pay more than $5 or $6 for a sufficient quantity of ore to make one ton of iron. Anthracite coal may be mined for various prices, ranging from 50 cents to $2, and upwards, per ton; bituminous coal, from 25 cents to $1 per ton. These prices are charged upon the gross ton, of 2240 pounds, as it is delivered at the mouth of the mine. Other minerals are paid for in proportion to their hardness, and the difficulty with which they are mined. Good undermining, and large rooms, have a great effect in reducing the cost of minerals. Superior tools and sound timber have also a beneficial effect. Water and impure air have a tendency to cause high-priced work, and should be obviated by all means.

THE END.

www.ingramcontent.com/pod-product-compliance
Lightning Source LLC
LaVergne TN
LVHW050528100826
845148LV00002B/481

9781425519650